PATHWAYS OF POLLUTANTS IN THE ATMOSPHERE

PATHWAYS OF POLLUTANTS IN THE ATMOSPHERE

A ROYAL SOCIETY DISCUSSION
ORGANIZED BY T. M. SUGDEN, F.R.S.,
FOR THE ROYAL SOCIETY'S STUDY GROUP
ON POLLUTION IN THE ATMOSPHERE

HELD ON 3 AND 4 NOVEMBER 1977

LONDON
THE ROYAL SOCIETY
1979

Printed in Great Britain for the Royal Society
at the
University Press, Cambridge

ISBN 0 85403 107 3

First published in *Philosophical Transactions of the Royal Society of London*, series A, volume 290 (no. 1376), pages 467–637

Published by the Royal Society
6 Carlton House Terrace, London SW1Y 5AG

PREFACE

Any material emitted into the atmosphere follows pathways along which chemical change may occur, before it or its products reach some form of sink at which it or they are removed, forming a part of the total biogeochemical cycles of the elements. Some substances are emitted as waste products of industrial or other human activities, and are recognized as pollutants by the damage they can inflict on a variety of targets, especially, but by no means uniquely, human ones. Damage may be direct or, if resulting from effects of pollutants on climate, indirect.

Two broad areas may be distinguished in the scientific study of pollution, these being respectively the establishment of reliable dose–effect data, and the pattern of exposure of targets to known dosage, current or foreseen. The latter requires adequate knowledge of the distribution of pollutants along their pathways in the atmosphere from sources to sinks; it implies effective budgeting of pollutants and the placing of such budgets in the context of natural ones.

This volume contains papers presented at a two-day Discussion Meeting organized through the Society's Study Group on Pollution in the Atmosphere, and held in November 1977 in the rooms of the Royal Society. Most of these papers deal with selected problems in the elucidation of atmospheric pathways for a number of important pollutants on a local, regional or global scale, involving atmospheric physics and chemistry as well as general meteorology in the lower atmosphere and the stratosphere. They were invited for the Discussion because they provide at the same time accounts of current work on particular problems, and exemplifications of the various approaches used generally in measurements and models of pollution in the atmosphere.

The ultimate objective of scientific research about pollution lies in the need to formulate policies for control, which to make sense require a rational basis for the assessment of risk. Consequently the later papers in the volume deal with impacts on targets, with the monitoring of pollution, and with the approach adopted in the United Kingdom to control policy.

Past and present staff of the Royal Society must receive generous thanks in recognition of their skilful organization of the Discussion, especially Mrs P. B. Carter and Miss C. A. Johnson, as must also Dr M. B. Goatly for efficient handling of the editorial task.

October 1978 T. M. SUGDEN

CONTENTS

[One plate]

Phil. Trans. R. Soc. Lond. A. **290**, 469–476 (1979)
Printed in Great Britain

The classification of pollutants and their pathways in the atmosphere

By T. M. Sugden, F.R.S.
Trinity Hall, University of Cambridge, CB2 1TJ, U.K.

Atmospheric pollutants may damage, directly or indirectly, human life and health, other living organisms and complete ecosystems, human artefacts, and climatic conditions. The development of appropriate policies and methods for control of pollution requires, *inter alia*, an assessment of the routes taken by pollutants or their precursors through the atmosphere. Consideration of these routes leads to a broad classification on a local, regional or global basis associated mainly, but not exclusively, with the terrestrial boundary layer, with the troposphere and with the stratosphere respectively. This may require in some cases the perspective of total biogeochemical cycles, and in any event of the relative importance of man-made and natural sources of materials to be regarded as pollutants.

1. Introduction

Pollution is not a branch of natural philosophy but a social phenomenon to which a scientific approach is mandatory if effective methods of dealing with it are to be found and applied. To pollute, according to Dr Johnson, means 'to corrupt by mixture of ill, either moral or physical', a definition confirmed in the latter, presently relevant sense by the *Oxford English Dictionary* as 'to make physically impure, foul or filthy: to dirty, stain, taint, befoul'. A long-standing result of human activities is the release into the atmosphere of waste products, either accidentally or deliberately, which do just these things with actual or potential consequences of damage.

These consequences may be itemized broadly as follows:

(1) damage to human life or health;

(2) damage to other living organisms, including crops and extending to ecosystems in general;

(3) damage to buildings and other artefacts;

(4) modifications to climate with all the consequential changes which could occur falling under the first three headings.

Two trends have become very apparent in recent years. The first is the vastly increased use of new technologies and associated inroads on the material resources of the planet, with the consequences that there are much greater volumes of waste product, in greater variety, thereby resulting in larger pollution hazards and newer ones. Secondly, social awareness has developed in such a way as to find these hazards variously unacceptable, and to seek to control them by minimizing or eliminating them. The impact of developing technology and the concern about its implications for pollution are, of course, not uniform over the world.

The control of pollution is to be visualized as taking place within a framework of legislation which permits the encouragement of restraint and the curbing of excess. Any such framework implies a philosophy of control and corresponding policies. This is not the place to undertake a review of such matters but current concern with them in the United Kingdom is emphasized by the appearance of two recent reports; one on pollution control in Great Britain and how it works (Department of the Environment 1976*a*); the other on an integrated approach to air-pollution control (Royal Commission on Environmental Pollution 1976) in which recommendations for changes in present methods of handling some of the problems are put forward.

The general issue of control is raised at the outset in order to emphasize that the study of pollution is a practical matter, and that the fundamental objective of any scientific investigation of it must be to assist decision-makers by providing them with the best available information. Such information comprises, for the most part:

(1) estimates on a quantified basis of the actual or potential damage caused by specific pollutants, or combinations of them;

(2) the identification of existing and potential sources of pollutants, and the placing of them in perspective against natural sources;

(3) the relation of sources to effects.

The last of these implies a knowledge of the pathways of pollutants, from identified sources to identified sinks. Together these kinds of information enable an assessment of the total risk situation and the identification of areas where control can best be applied. Further information falls under the headings of practical suggestions for methods of control, and programmes for research into pollution affairs. Authorities may then set up cost/benefit analyses on whatever basis they decide may be appropriate in order to decide on action.

2. The Royal Society Study Group on Pollution in the Atmosphere

The Royal Society Study Group on Pollution in the Atmosphere was set up in 1975 with the principal terms of reference of undertaking an intensive study of the long-range dispersal, absorption and chemistry of pollutants in the atmosphere and their effect on the environment, including man-made ecosystems; and to arrange for a Discussion Meeting on this subject in the 1976/77 session; it is that Discussion Meeting which has given rise to the present series of papers.

The Group was made up of about 25 individuals drawn from universities, government departments and industry, and represented the wide variety of scientific disciplines involved – physics, chemistry, meteorology, chemical engineering, fluid mechanics and biology. It has maintained a working interface with a parallel Study Group on Long Term Toxic Effects – mostly in humans – which held a similar Discussion Meeting shortly after.

The Group held its first scientific meeting in September 1975, and since then has had a further 12 such meetings. At a very early stage it adopted a method of approach which was essentially that of the present Discussion – that of pathways of pollutants. Early meetings were given over to general considerations of the transport of matter through the atmosphere, of the kinds of chemical change which are important during that transport, and of deposition or absorption from the atmosphere, drawing examples from the behaviour of various pollutants. This laid a foundation within which it was possible to discuss particular problems relating to the total behaviour of specific pollutants. Meetings were held on the impact of supersonic flight on the stratosphere, the impact of fluorocarbons, the transport of sulphur compounds in northwest Europe, airborne heavy metals, carbon dioxide in the atmosphere, and the philosophy and practice of air pollution monitoring. Finally, the Group gave consideration to transient pollution: that resulting from accidental release into the atmosphere of more or less toxic substances.

3. Pathways in pollution: general

The total amount of a pollutant released depends on the number and size of its sources. Its capability of doing damage is conditioned by the amount of it available at the place where it is taken up from the atmosphere. Consequently it becomes important to know the pattern of its dispersal from the original sources, the way in which chemical change may occur during its migration, and the efficiency of the sinks through which it is finally removed from the sphere of effective action. That action may be of course be direct, as in inhalation or in deposition on the surface of an organism or material which as a result suffers harm; or it may be less direct, as for example with acidified rain which may induce changes in the character of soils or watercourses to which it penetrates.

The shortest known pathway of atmospheric pollution is almost certainly that between the burning tip of a cigarette and the lungs of the smoker. It has been treated extensively elsewhere and is outside the present terms of reference. Similarly, indoor pathways have been excluded, although the extensive work that has gone into the rigorous control of atmospheres in places of work where toxic substances may be a hazard has produced many results that are of value in the setting of standards in wider contexts. The quality of domestic indoor atmospheres may sometimes leave much to be desired but is not of great interest in considering atmospheric pollution in general.

The atmosphere proper which is our concern may be divided roughly into three zones. The uppermost is the stratosphere, which extends to a height of some 50 km above the surface from about 8 km above the poles to about 16 km above the equator, subject to seasonal and other fluctuations. It is bounded by temperature inversions and is characterized by rising temperature with increasing height. The pressure at the lower bound is of the order of one-tenth that at the surface, and at the upper bound about one-thousandth of that pressure. It is generally calm because of its temperature gradient, which limits convection, free of cloud except for a little near its base, and free of precipitation. Regions above the stratosphere do not appear as yet to have produced issues important for pollution.

Below the stratosphere is the troposphere, extending downwards almost to the Earth's surface, and with a temperature gradient in the opposite direction. The upper interface, called the tropopause, is not readily crossed by diffusing gases, and our understanding of transfer across it is incomplete, especially in middle latitudes where it is ill defined, being spread over an extent of a few kilometres called the tropopause gap in which the temperature gradient is indeterminate. In terms of motion the troposphere is relatively violent, since it responds vigorously to the uneven heating of the land masses and the oceans with the production of winds. It is associated with cloudiness, often extreme, and with the precipitation of rain or snow. The troposphere is very important in determining general patterns of climate.

Close to the surface of the Earth the troposphere is modified in a boundary layer, very roughly defined, extending vertically a kilometre or so. In this region the atmosphere is responsive to detailed features of the surface such as mountains, valleys, coastal littorals, conurbations or even large individual buildings. At times its upper bound may be associated with a temperature inversion, providing a region from which the upward escape of air is difficult. A useful account of the physics of planetary atmospheres, and the terrestrial one in particular, has recently been given by Houghton (1977).

Sources of pollution are located almost exclusively at or very close to the surface of the Earth, aircraft providing the exception. By far the commonest pollutants are the combustion products of fossil fuels. The major products are carbon dioxide and water, the latter of which can hardly be regarded as a pollutant because of the much larger amount of water vapour deriving from natural sources. Carbon dioxide merits consideration, however, because the amount currently being emitted is sufficient to cause some increase in the total atmospheric content. Carbon dioxide is important in photosynthesis and in the absorption of infrared radiation. It is dealt with in one of the later papers in this volume. The combustion products also constitute the major source of man-made sulphur oxides, derived from sulphur compounds in the fuel, extreme refining of which is prohibitively expensive. They are further the major source of nitrogen oxides – nitric oxide and nitrogen dioxide – which are formed by the fixation of atmospheric nitrogen during the combustion process. Automotive fuels also yield unburnt hydrocarbons and related substances – some of which may be carcinogenic – as well as carbon monoxide and compounds deriving from anti-knock additives, predominantly of lead. Another product, black smoke, has largely been eliminated in recent years because of the application of official policies about industrial and domestic effluents.

Direct industrial emissions of other, potentially toxic materials are minimal because of careful control of stacks, as they are in general domestic emissions. A recent important exception to this is the recent widespread dissemination of chlorofluoromethanes from refrigeration devices and aerosol sprays. Agriculture provides sources of pollution. Denitrifying bacteria in the soil provide a source of nitrous oxide which may become a source of concern with the increasing use of nitrogenous fertilizers. Not all gaseous pollutants are associated with modern technology: primitive slash-and-burn agriculture contributes to the emission of chlorine compounds. Finally the possibility exists of the accidental discharge, in single incidents, of significant amounts of substances which may be of high toxicity, such as radioactive materials and industrial products and intermediates.

Particular pollutants may have many and varied pathways from source to sink, both in atmospheric trajectory and in chemical change along the path. Thus, nitrogen oxides released from an automobile exhaust may react, via rather complicated chemistry, with other emission products to give lachrymatory substances in aerosol form such as peroxyacetyl nitrate (PAN), which make up photochemical smog – a *local effect*. If emitted from a high stack, a proportion of such oxides may escape from the boundary layer into the troposphere generally, and appear as nitrous or nitric acids in precipitated rain several hundred miles away, in which case there is an example of *regional* pollution. Because of their reactivity or solubility they are most unlikely to reach the stratosphere. If they are actually emitted there, however, from the exhaust of a supersonic aircraft they will be disseminated round it fairly quickly (a few months) with small chance of escape downwards and only slow upwards migration; they will then engage in the complicated, partially cyclic sets of reactions which can result in a decrease in the amount of stratospheric ozone until they form nitric acid which eventually reaches the troposphere and is rained out. This provides an example of *global* pollution.

The practical object of any study of the pathways of a pollutant is the construction of as complete an inventory as may be possible of the substance concerned, so that the contribution of a particular source or group of sources to the various sinks can be estimated. This requires more or less extensive measurement or monitoring of the substance, as well as substances with which it may react in the atmosphere. It also tends to need the construction and application of an appropriate model of the situation.

As atmospheric pathways become longer in space and in time, the more blurred and difficult to unravel they tend to become. Part of this difficulty stems from the more extensive sampling and modelling required, the latter of which can be a burden on even large computer systems. Part of it arises because the paths themselves intersect and overlap. Increasingly it becomes necessary to take into account a wider range of sources, including natural ones. Ultimately there may be significant impact on the total biogeochemical cycle of one or more elements. In this respect the work which is being done under SCOPE auspices (Svensson & Soderlund 1976) on the global cycles of nitrogen and sulphur is of great importance, as well as work currently proceeding on the carbon cycle.

4. Local pollution

In the local situation, sources are usually clearly defined and are almost exclusively man-made. Pathways in the atmosphere lend themselves to representation by sophisticated techniques of modelling, and monitoring is a relatively straightforward affair. A typical very localized situation is the deposition of lead compounds from automobiles close to highways and road junctions. Materials emitted from high stacks can be dealt with by using modelling techniques associated with the cases of representative weather conditions proposed by Pasquill (1961) which allow estimates to be made of the amount of pollutant retained in the boundary layer and the amount taken up into the troposphere. Effective times from emission to deposition range from a few minutes to a day or two. This applies to both steady or intermittent emissions and to near instantaneous accidential releases.

Much more remains to be found out about deposition itself, and about the subsequent pathways in targets of pollution to the point where actual damage is induced. The hazards to people, vegetation or artefacts are not easy to estimate because of the problems of getting valid information about the effects of long duration of exposure at low levels of concentration. The situation is quite different when dealing with a single accidental release of quantities of a substance which may be acutely toxic. The techniques of control are also quite different, being related to design to prevent mechanical failure of plant rather than with the establishment of permissible low levels of emission.

5. Regional pollution

Regional pollution, occurring at up to a few hundred miles from sources, is rare because of the relatively wide spread from the source and consequent high dilution of the pollutants. For a given type of environment the local effects would be expected to much more severe and demanding of attention. The exception is found where for some reason the remoter target is particularly vulnerable.

The most characteristic example of this is the deposition of acid rain, deriving from the sulphur and nitrogen oxides of combustion products, falling on soils which are weakly buffered. This can result in damage to vegetation, and to animal life in rivers and lakes which become too acid.

The analysis of pathways in such cases tends to be a matter of great complication in part because of the wide range of sources of pollution which can contribute and the possible intervention of effects from natural sources. A further source of complication lies in the complexity of the atmospheric trajectories, which take a few days to traverse, in the rather violent motions of the troposphere. Extensive monitoring, including aerial observations, must be used in conjunction with the modelling. A recent study in this category is the intensive work sponsored by O.E.C.D. on the transport of sulphur compounds across North-West Europe (see, for example, Department of the Environment 1976*b*).

6. Global pollution

Global pollution is largely synonymous with stratospheric pollution. Mixing times there are in the region of a few weeks in the E–W dimension, a few months in the N–S dimension, while vertical motion is generally much slower in the lower parts of the stratosphere, with characteristic times of a few years. Transfer to and from the troposphere and between hemispheres is also slow. The result is a global, or at least hemispheric distribution of pollutants. Chemically stable substances which are not readily rained out of the troposphere can also achieve world-wide distribution.

The pathways of stratospheric pollution start from two kinds of source: direct injection into the stratosphere and penetration of the tropopause by substances released at a lower level which are stable enough to survive upward transfer from the surface. Attention has been focused very heavily on the effects of pollutants on the ozone in the stratosphere because of its efficiency in filtering out potentially harmful ultraviolet radiation.

In the upper parts of the stratosphere, radiation of wavelength below about 220 nm is removed by absorption by diatomic oxygen molecules, which are in consequence dissociated to free atoms. These atoms can combine with further diatomic oxygen to form ozone (O_3), which can then be destroyed by a variety of further reactions. The result is to give a near steady state amount of ozone. Ozone itself is a very effective absorber for radiation up to a wavelength of about 290 nm, its effectiveness then falling rapidly to zero at about 320 nm. It is the absorption of this radiation and its ultimate degradation to heat which is largely responsible for the temperature distribution of the stratosphere (and therefore for the stratosphere itself).

The narrow band of radiation between 290 and 320 nm, known as u.v.-B, appears to be particularly biologically active. This is perhaps not very surprising in that the quanta concerned are equivalent to about 400 kJ mol^{-1}, which is near the threshold of breaking many of the chemical bonds found in organic materials. There is, for example, a damaging action on DNA. More especially, there is a significant correlation between exposure to u.v.-B and non-melanoma skin cancer, a relatively frequent condition of low mortality to which some groups of white-skinned people are particularly prone. There is also evidence that some cases of the much rarer, but much more dangerous, melanoma skin cancer are induced by exposure to u.v.-B.

Absorption of u.v.-B by ozone in the stratosphere is such that a decrease of 1 % in total ozone lets through about 2 % more u.v.-B, with consequent and not necessarily linear effects on skin cancer incidence; hence the major part of the current concern. There may also be consequences for other animals and for vegetation, etc., but little useful evidence is available. A sufficient change in ozone and its distribution might also affect the atmospheric heat balance in ways which are difficult to predict with accuracy.

Anxiety has stemmed from three main types of source of pollution:

(1) the injection of nitrogen oxides directly into the stratosphere from the exhausts of supersonic aircraft;

(2) the penetration from below of the very stable chlorofluoromethanes, used as refrigerants and aerosol sprays, in increasing amounts;

(3) the penetration of nitrous oxide, released by denitrifying bacteria, and its potential increase through increased use of nitrogenous fertilizers.

All of these provide sources of materials which can take part in an extensive and complicated scheme of chemical reactions in which ozone is concerned, and which can result in a net decrease of total ozone. There have been very extensive and sophisticated modelling studies of the stratospheric oxone situation, which is a difficult one because of the linking of the chemistry with transport in the stratosphere and because of uncertainties about whether the chemistry is complete and about the values of some critical rate constants.

The supersonic flight problem is probably the most tractable in that direct injection will produce relatively immediate effects – within a few years. The predictions made in a National Academy of Sciences Report (Booker 1975), given the expected low rate of growth of supersonic transport, give no cause for alarm in that ample time is available for improvements in engine design to limit emissions.

The other two sources listed above provide more serious problems because of the size of the potential effects and of the delay in their appearance associated with the delays in the penetration to the stratosphere of the original emissions. To quote from a recent report (Tukey 1976) on the environmental effects of chlorofluoromethane (CFM) release: 'If CFM uses were to continue at a constant rate, the ozone reduction and consequent u.v.-B increase would flatten out gradually, approaching a steady state. To reach half of this value would take roughly 50 years.'

Consequently one may be concerned about the later developments from pollutants which may already be launched. Fortunately, the work which has already been done (Gutowsky 1976), within 4 years of the recognition of the problem, indicates that there is here effectively an early warning situation: 'It appears that the continued release of CFMs at the 1973 production rate would cause the ozone to decrease steadily until a probable reduction of about 6–7.5 % is reached, with an uncertainty range of at least 2–20 %, using what are believed to be roughly 95 % confidence limits.' Given the time scale and the potential magnitude of the effect there is thus a period of a few years which will enable production to be eased, less potentially harmful substitutes sought, and the predictions sharpened up by improved modelling coupled with direct observations.

It deserves to be emphasized that it is u.v.-B which is likely to have the harmful consequences, and the amount of this which reaches the surface of the Earth now is some ten times greater than the increase which would result from a decrease of 5 % in stratospheric ozone. This could point up the need for a present effective campaign against exposure to sunlight in appropriate areas, of which there is little indication.

References (Sugden)

Booker, H. G. 1975 *Environmental impact of stratospheric flight.* Washington, D. C.: Climatic Impact Committee, National Academy of Sciences.

Department of the Environment 1976*a* *Pollution control in Great Britain: how it works* (Pollution Paper No. 9). London: H.M.S.O.

Department of the Environment 1976*b* *Effects of Airborne Sulphur Compounds on Forests and Freshwaters* (Pollution Paper No. 7). London H.M.S.O.

Gutowsky, H. S. 1976 *Halocarbons: effects on stratospheric ozone.* Washington, D.C.: Panel on Atmospheric Chemistry of the Committee on Impacts of Stratospheric Change, National Academy of Sciences.

Houghton, J. T. 1977 *The physics of atmospheres.* Cambridge University Press.

Pasquill, F. 1961 *Met. Mag.* **90**, 33.

Royal Commission on Environmental Pollution 1977 *Air pollution control: an integrated approach* (Fifth Report, Cmnd 6371). London: H.M.S.O.

Svensson, B. H. & Soderlund, R. 1976 *Nitrogen, phosphorus and sulphur – global cycles.* SCOPE Report 7, Ecological Bulletin. Stockholm: Swedish National Research Council.

Tukey, J. W. 1976 *Environmental effects of chlorofluoromethane release.* Washington, D.C.: Committee on Impacts of Stratospheric Change, National Academy of Science.

Discussion

S. W. RADCLIFFE (*Room 708 B, Scientific Branch, Greater London Council, The County Hall, London SE1 7PB, U.K.*). In the early part of his paper Dr Sugden has classified the effects of pollution according to physical damage to living organisms, material and climate. I think that the changes which pollution can make to amenity should also be included. For about 20 years the inhabitants of London and other large cities have been enjoying the cleaner air brought about by smoke control. I believe this would rank in their minds as just as great a benefit as any mitigation of physical damage. Airborne pollutants that have important amenity effects are still common nowadays; there are, as examples, the stench and dirtiness of road traffic, the smells of food processing and chemical plant, and the noise and vibration of traffic and aircraft. Amenity effects are even more difficult to deal with quantitatively than physical effects, and are therefore possibly less attractive subjects for scientific investigation. I feel, however, that they should always be kept in mind in general discussions about pollution and its control.

T. M. SUGDEN. Many of the benefits stemming from smoke control in London and other large cities are directly physical in nature, both in improved health of people and in reduction of damage to artefacts. I agree that amenity aspects of atmospheric pollution are important: in general they are local in nature, arise from well-known sources and therefore in principle are not difficult to control.

B. A. THRUSH, F.R.S. (*University of Cambridge, Department of Physical Chemistry, Lensfield Road, Cambridge CB2 1EP, U.K.*). One positive aspect of the release of anthropogenic compounds into the atmosphere is that, unlike natural substances, they can provide atmospheric tracers with accurately known source strengths. A familiar example is the rate of redistribution of chlorofluoromethanes, which are released mainly in northern temperate latitudes. The determination of the atmospheric lifetime of methyl chloroform probably provides the most straightforward method for determining the globally averaged tropospheric concentration of hydroxyl radicals, which are believed to provide the main atmospheric sink for many hydrocarbon derivatives.

Phil. Trans. R. Soc. Lond. A. **290**, 477–494 (1979)
Printed in Great Britain

A comparison of one-, two- and three-dimensional model representations of stratospheric gases

By A. F. Tuck
Meteorological Office, London Road, Bracknell, RG12 2SZ, U.K.

Three- and two-dimensional model results have been averaged to investigate conceptual errors in two- and one-dimensional models. Average dynamical quantities show inter-hemispheric asymmetries in both mean and eddy vertical motions, with anomalous behaviour of tracers near effective source and sink regions.

Zonal, hemispheric and global means of the rates of gas reactions show large deviations between terms like $\overline{k[\mathrm{A}][\mathrm{B}]}$ and $\bar{k}\overline{[\mathrm{A}]}\,\overline{[\mathrm{B}]}$, causing significant errors in two- and one-dimensional model calculations. These errors are often associated with dynamical features such as jet streams or the tropopause, and affect the entire model atmospheres except in the summer mid-stratosphere.

It is concluded that correlated measurements of atmospheric molecular number densities are urgently required to understand the deficiencies in models, which have been widely used to make perturbation calculations of the effects of aircraft and chlorofluoromethanes on stratospheric ozone. The sources of error described in this work arise from inadequacies in the formulation of one- and two-dimensional models, rather than from uncertainties in the input data, and have not been included in published error analyses.

1. Introduction

In order to calculate the chemical composition of the stratosphere, and possible perturbations to its distributions of molecules, it has been necessary to resort to numerical models in one, two and three dimensions.

A three-dimensional general circulation model (g.c.m.) of the atmosphere is based on the equations of motion for the wind components, the mass continuity equation, the equation of state and the thermodynamic equation. Integration of this set of equations with time yields values of the wind components, pressure, density and temperature on a grid of longitude, latitude and altitude. The rate of change of any gaseous constituent is given by an equation of the type

$$\partial \chi_i / \partial t = P_i - L_i - \nabla \cdot \boldsymbol{V} \chi_i, \tag{1}$$

where χ_i is mixing ratio, P_i is the local rate of production (source), L_i is the local rate of loss (sink) and the last term is the flux divergence caused by the wind velocity $\boldsymbol{V}$.

In a g.c.m. including the stratosphere it might be necessary to include up to 60 species for a complete representation of the photochemistry. In addition many physical processes such as radiative transfer, surface exchanges, subgrid scale convection, rainfall and the like must be included. This is not possible even with the largest and fastest machines presently available, and in practice no more than three or four species have been represented by an equation like (1).

Two-dimensional models economize on the number of grid points by averaging over longitude (i.e. round a latitude circle) and representing the variables as zonal means on a grid of latitude and altitude. One-dimensional models carry this process a stage further by averaging over latitude to represent the atmospheric variables in a single vertical column.

It will be seen in § 2 that this averaging process raises questions about how accurately models of reduced dimensionality represent true zonal or global means of three-dimensional distributions of variables. The work described here seeks to investigate this question by using results from Meteorological Office models. Mahlman (1976) has also considered these problems.

There is beyond this, of course, the more fundamental question of whether models of one, two or three dimensions mishandle correlations of variables occurring in the real atmosphere on scales not represented in the models. For example, it is likely that ascending air over the continents has a different chemical composition from that of descending air over the oceans. At any one time these air parcels cannot of course react chemically; however, such a distinction can only be made in a three-dimensional model, and then only on scales larger than its grid length. It thus follows that if it is desired, for example, to compute the mean global rate of a reaction

$$A+B \rightarrow C+D$$

characterized by rate coefficient k, it is in principle necessary to compute from observed number density distributions of A and B, pressure and temperature the local values of the rate k[A] [B], and obtain the global mean rate by averaging these local rates. This mean will be different from the mean computed by forming the product of the separate global means of k, [A] and [B], if any nonlinear correlations are present. The computation of zonal means in two-dimensional models and of global means in one-dimensional models is implicitly dependent upon the assumption that the true mean rate can be formed by the product of the means of k, [A] and [B].

The necessary observations of molecular number densities are not available, so the checking of the averaging process in one- and two-dimensional models unfortunately cannot be made with real data. However, it is possible for example to average data from a three-dimensional model to check both two- and one-dimensional calculations, and to average two-dimensional model results to check one-dimensional models. Questions as to the effect of averaging over different seasons and the two hemispheres may also be examined as part of this approach. The work described here has attempted to carry out such an investigation, and gives first results.

2. Equations and models

In three dimensions the behaviour of the mixing ratio χ of a constituent gas is given by

$$\frac{\partial \chi}{\partial t}+\frac{1}{r\cos\phi}\frac{\partial}{\partial \Lambda}u\chi+\frac{1}{r\cos\phi}\frac{\partial}{\partial \phi}v\chi\cos\phi+\frac{\partial}{\partial z}w\chi = P-L. \tag{2}$$

Here P is source, L is sink, u, v and w are respectively zonal, meridional and vertical velocities, ϕ is latitude, Λ is longitude, z is altitude and t is time.

Reduction to two dimensions is achieved by the definition of the zonal mean of χ (denoted by an overbar)

$$\overline{\chi} = \frac{1}{2\pi}\int_0^{2\pi} \chi\, \mathrm{d}\Lambda. \tag{3}$$

This results in a relation which describes the time dependence of χ in the meridional and vertical directions:

$$\frac{\partial \overline{\chi}}{\partial t}+\frac{1}{r\cos\phi}\frac{\partial}{\partial \phi}\overline{v\chi}\cos\phi+\frac{\partial}{\partial z}\overline{w\chi} = \overline{P-L}. \tag{4}$$

An equation like (4) exists for each species and should in principle be integrated forward in time, with the dynamical terms on the left and the physico-chemical terms on the right evolving simultaneously and interactively on the latitude–height grid. However, a full two-dimensional set of the equations of motion, thermodynamic equation and a continuity equation for each species demands in its solution a general theory connecting the eddy components with the mean components of u, v, w, χ and temperature T. This task has not in general been tackled by two-dimensional modellers. Instead an approach given originally by Reed & German (1965) has been widely adopted, in which climatological data are used for the mean meridional wind component v in conjunction with a flux-gradient representation of the eddy terms derived from empirically obtained diffusivity coefficients.

The zonal mean $\bar{\chi}$ of equation (3) is used to define the local departures χ' from the average at any longitude:

$$\chi' = \chi - \bar{\chi}. \tag{5}$$

The zonal mean of χ' is zero, but products of primed quantities lead to the relations

$$\overline{v\chi} = \bar{v}\bar{\chi} + \overline{v'\chi'}; \tag{6}$$

$$\overline{w\chi} = \bar{w}\bar{\chi} + \overline{w'\chi'}. \tag{7}$$

The term on the left side of (6) and (7) is the mean total flux, and is equal to the sum of the flux $\bar{v}\bar{\chi}$ or $\bar{w}\bar{\chi}$ caused by the mean components and the eddy flux $\overline{v'\chi'}$ or $\overline{w'\chi'}$ arising from correlations between local departures from zonal averages at different longitudes.

The Reed & German (1965) approach results in expressions of the form

$$\overline{v'\chi'} = -\left(K_{yy}\frac{\partial\bar{\chi}}{\partial y} + K_{yz}\frac{\partial\bar{\chi}}{\partial z}\right); \tag{8}$$

$$\overline{w'\chi'} = -\left(K_{zy}\frac{\partial\bar{\chi}}{\partial y} + K_{zz}\frac{\partial\bar{\chi}}{\partial z}\right). \tag{9}$$

The theory requires K_{yy} and K_{zz} to be positive, but K_{yz} ($\approx K_{zy}$) may be of either sign. Methods of obtaining these coefficients have been discussed in various places (Reed & German 1965; Gudiksen *et al.* 1968; Luther 1973). The further averaging of the two-dimensional model over latitude to produce a one-dimensional model immediately produces a dilemma as to whether the resulting column should be chosen to represent global or hemispheric behaviour. The derivation of global and hemispheric models will be given in parallel.

To obtain one-dimensional equations from equation (4), it is necessary to define global/hemispheric averages by the following relations. Globally,

$$\langle\chi\rangle = \frac{1}{2}\int_{-\frac{1}{2}\pi}^{\frac{1}{2}\pi} \bar{\chi}\cos\phi\,\mathrm{d}\phi; \tag{10a}$$

hemispherically,

$$\langle\langle\chi\rangle\rangle = \int_{0}^{\frac{1}{2}\pi} \bar{\chi}\cos\phi\,\mathrm{d}\phi. \tag{10b}$$

Here, the single angle brackets denote a global mean and the double angle brackets a hemispheric mean. These lead respectively to one-dimensional equations

$$\frac{\partial\langle\chi\rangle}{\partial t} + \frac{\partial}{\partial z}\langle w\chi\rangle = \langle P-L\rangle, \tag{11a}$$

$$\frac{\partial\langle\langle\chi\rangle\rangle}{\partial t} + \frac{(\overline{v\chi})_{\text{equator}}}{r} + \frac{\partial}{\partial z}\langle\langle w\chi\rangle\rangle = \langle\langle P-L\rangle\rangle. \tag{11b}$$

Equation $10a$ for air ($\chi = 1$) would lead to a latitude of 0° for the global average column, and ($10b$) would lead to 30° for the hemispheric average column. For species with χ much less than unity (i.e. all species except N_2 and O_2), the latitude should be at the centre of gravity of $c \cos \phi$ (Tuck 1977), with the difficulty that this may occur at different latitudes for different species. The global mean vertical velocity must be zero in the temporal mean, and hence ($11a$) is used in the form

$$\frac{\partial\langle\chi\rangle}{\partial t} = \frac{\partial\langle K_z\rangle}{\partial z}\frac{\partial\langle\chi\rangle}{\partial z} + \langle P - L\rangle, \tag{12}$$

where the global mean coefficient $\langle K_z\rangle$ is defined by

$$\langle w\chi\rangle = -\langle K_z\rangle\,\partial\langle\chi\rangle/\partial z \tag{13}$$

and represents total flux, making no distinction between mean and eddy motion. Equation (12) implies that total global mean vertical transport of a species is proportional to and down its gradient, and suffers further from the difficulty that air motions require the mean latitude to be 0°, while it is clear that if the insolation for the photochemistry is calculated at this latitude the resulting solar fluxes will not represent a global average. Any effects such as seasonal differences or hemispheric asymmetries are averaged out.

A hemispheric equation analogous to (12) may be obtained from ($11b$):

$$\frac{\partial\langle\langle\chi\rangle\rangle}{\partial t} = \frac{\partial}{\partial z}\langle\langle K_z\rangle\rangle\frac{\partial\langle\langle\chi\rangle\rangle}{\partial z} - \frac{(\overline{v\chi})_{\mathrm{equator}}}{r} - \frac{\partial}{\partial z}(\langle\langle w\rangle\rangle\langle\langle\chi\rangle\rangle) + \langle\langle P - L\rangle\rangle, \tag{14}$$

where the hemispheric vertical flux is resolved into mean and eddy components:

$$\langle\langle w\chi\rangle\rangle = -\langle\langle K_z\rangle\rangle\,\partial\langle\langle\chi\rangle\rangle/\partial z + \langle\langle w\rangle\rangle\langle\langle\chi\rangle\rangle. \tag{15}$$

If it is desired to investigate such effects as seasonal variations, hemispherical differences and the relative importance of mean and eddy motions in the vertical, a one-dimensional model must be based on (14) rather than (12). In general, hemispheric mean vertical velocities would be extremely difficult to compute from real data; here, three-dimensional model data have been used.

The practical formulation of a one-dimensional model from (14) involves rewriting the two-dimensional equation (4) in terms of number densities:

$$\frac{\partial\bar{n}}{\partial t} + \frac{1}{r}\frac{\partial(\overline{nv})}{\partial\phi} - \overline{nv}\frac{\tan\phi}{r} + \frac{\partial(\overline{nw})}{\partial z} = \overline{P^{\dagger} - L^{\dagger}}, \tag{16}$$

where $P^{\dagger}$ and $L^{\dagger}$ now refer to number densities rather than mixing ratios.

By writing $y = r\phi$, and assuming that mass continuity relates $\bar{v}$ and $\bar{w}$ it may be shown that to a good approximation

$$\frac{\partial\bar{v}}{\partial y} - \bar{v}\frac{\tan\phi}{r} + \frac{\partial\bar{w}}{\partial z} = 0. \tag{17}$$

This relation makes it clear that a one-dimensional model representing a hemisphere must have a column in the opposite hemisphere from which to supply horizontal flow to satisfy mass continuity when the hemispheric mean vertical velocity is non-zero. It is thus in a sense two-dimensional; it may be made effectively one-dimensional by treating one of the columns as a boundary, with prescribed values.

From (16),

$$\frac{\partial\bar{n}}{\partial t} + \frac{\partial(\overline{nv})}{\partial y} - \overline{nv}\frac{\tan\phi}{r} + \frac{\partial(\overline{nw})}{\partial z} = \overline{P^{\dagger} - L^{\dagger}}. \tag{18}$$

By using (6) to (9) in number density form, differentiating products and noting (17),

$$\frac{\partial \bar{n}}{\partial t}+\bar{w}\frac{\partial \bar{n}}{\partial z}-\frac{\partial}{\partial z}\rho\left\{K_{zy}\frac{\partial}{\partial y}\frac{\bar{n}}{\rho}+K_{zz}\frac{\partial}{\partial z}\frac{\bar{n}}{\rho}\right\}+\bar{v}\frac{\partial \bar{n}}{\partial y}-\frac{\partial}{\partial y}\rho\left\{K_{yy}\frac{\partial}{\partial y}\frac{\bar{n}}{\rho}+K_{yz}\frac{\partial}{\partial z}\frac{\bar{n}}{\rho}\right\}$$
$$+\rho\left\{K_{yy}\frac{\partial}{\partial y}\frac{\bar{n}}{\rho}+K_{yz}\frac{\partial}{\partial z}\frac{\bar{n}}{\rho}\right\}\frac{\tan\phi}{r}=\overline{P^{\dagger}-L^{\dagger}}, \qquad (19)$$

where ρ is the air density.

The last four terms on the left side of (19) specify the mean and eddy components of the inter-hemispheric flow necessary to satisfy mass continuity. The remaining terms may be averaged by using (10*b*) to obtain the hemispheric one-dimensional form. In practice, hemispheric mean values, $\langle\langle w\rangle\rangle$, have been obtained from the Meteorological Office three-dimensional model (Comesa 1975; O'Neill & Newson 1977) on 13 levels up to 44 km, together with values of $\langle\langle K_z\rangle\rangle$ for ozone, water vapour and aircraft-emitted NO_x, which apart from prescribed sources and sinks, were treated as tracers.

The right side of equation (2), the term $\overline{P^{\dagger}-L^{\dagger}}$, is the net source term. In practice it includes fluxes from the planetary surface, losses due to rainout and the like, but in the upper stratosphere may be dominated by terms from chemical kinetic rate equations.

If the aircraft-emitted NO_x in the three-dimensional model is considered for the purpose of this exercise to be NO (there was no photochemistry in the model), the local value of the rate of the reaction

$$NO+O_3 \longrightarrow NO_2+O_2;\ k = 9.5\times 10^{-13}\exp\left(-1230/T\right)\,cm^3\,s^{-1}$$

may be computed, at any specified time, for each grid point. Averaging these local rates (which will be functions of the local values of [NO], $[O_3]$ and T) over longitude will give zonal mean rates at any level. This true zonal mean rate,

$$\bar{P}_{true} = \overline{k[NO]\,[O_3]}, \qquad (21)$$

may be compared with a zonal mean rate calculated by obtaining the product of the zonal means $\bar{k}$, $\overline{[NO]}$ and $\overline{[O_3]}$:

$$\bar{P}_{2D} = \bar{k}\overline{[NO]}\,\overline{[O_3]}. \qquad (22)$$

If there are significant differences between the values of $\bar{P}$ given by (21) and (22), then two-dimensional models are limited in accuracy to that extent in evaluating the rate of the reaction since they implicitly use (22).

A similar exercise was performed with two-dimensional model results, using (10*a*) and (10*b*), in order to examine the validity of one-dimensional evaluations of averaged rates. The three-dimensional model data were also further averaged down to true hemispheric mean and global mean values for the reaction between NO and O_3.

The reaction mechanism and rate coefficients used in the two-dimensional model are given in table 1.

The two-dimensional results used in this work have come from the Meteorological Office model described originally in Comesa (1975) and later updated and improved; see Murgatroyd (1977). It will suffice to say here that the model is formulated in 12 columns at 15° intervals from pole to pole and has 24 levels from surface to 48 km. The transport is in terms of $\bar{v}, \bar{\omega}(=dp/dt)$, where p is pressure, and the eddy diffusion coefficient components K_{yy}, K_{yz} and K_{zz}. The mean circulation data derive from climatological data up to about 30 km in the Northern Hemisphere, and up to 16 km in the Southern Hemisphere (see, for example, Newell *et al.* 1974) and from three-dimensional model calculations at higher levels. $\bar{\omega}$ was obtained from $\bar{v}$ by integrating the mass continuity equation downwards from the top boundary, where $\bar{\omega} = 0$. The values for K_{yy}, K_{yz} and K_{zz} are based on those of Luther (1973) but somewhat larger values were used in the Northern Hemisphere winter, and smaller values in the Southern Hemisphere winter. The pattern of variance in $\bar{v}$ over a year's run from the Meteorological Office three-dimensional model (COMESA 1975; O'Neill & Newson 1977) was used as a guide for modifications to the K's.

TABLE 1. CHEMICAL KINETIC MECHANISM, TWO-DIMENSIONAL MODEL

			$k/(\text{cm}^3 \text{ molecule s units})$
1.	$O_2 + h\nu$	$\longrightarrow O + O$	photodissociation rates diurnally and seasonally time-dependent
2.	$O_3 + h\nu$	$\longrightarrow O_2 + O(^1D)$	
3.	$O_3 + h\nu$	$\longrightarrow O + O_2$	
4.	$NO_2 + h\nu$	$\longrightarrow NO + O$	
5.	$HNO_3 + h\nu$	$\longrightarrow OH + NO_2$	
6.	$NO + h\nu$	$\longrightarrow N + O$	
7.	$O(^1D) + M$	$\longrightarrow O + M$	3.2×10^{-11}
8.	$O + O_2 + M$	$\longrightarrow O_3 + M$	$1.1 \times 10^{-34} \exp(507/T)$
9.	$O + O_3$	$\longrightarrow O_2 + O_2$	$1.9 \times 10^{-11} \exp(-2300/T)$
10.	$NO + O_3$	$\longrightarrow NO_2 + O_2$	$9.5 \times 10^{-13} \exp(-1230/T)$
11.	$NO_2 + O$	$\longrightarrow NO + O_2$	9.1×10^{-12}
12.†	$O(^1D) + N_2O$	$\longrightarrow NO + NO$	5.5×10^{-11}
13.‡	$OH + NO_2(+M)$	$\longrightarrow HNO_3(+M)$	k_{13}
14.	$OH + O_3$	$\longrightarrow HO_2 + O_2$	$1.6 \times 10^{-12} \exp(-1000/T)$
15.	$O(^1D) + H_2O$	$\longrightarrow OH + OH$	1.75×10^{-10}
16.	$O + HO_2$	$\longrightarrow OH + O_2$	4.3×10^{-11}
17.	$N + NO$	$\longrightarrow N_2 + O$	2.7×10^{-11}
18.	$N + O_2$	$\longrightarrow NO + O$	$1.1 \times 10^{-14} T \exp(-3150/T)$
19.	$N + O_3$	$\longrightarrow NO + O_2$	5.7×10^{-13}
20.	$OH + HO_2$	$\longrightarrow H_2O + O_2$	2.0×10^{-11}

† Other channel to $N_2 + O_2$ allowed for at same rate.
‡ $k_{13} = k_a k_b[M]/(k_a + k_b[M])$; $k_a = (68 - 0.22T) \times 10^{-12}$; $k_b = 3.7 \times 10^{-32} \exp(1093/T)$.

The averaging of two-dimensional model dynamical representations to obtain one-dimensional dynamical parameters may be considered (Murgatroyd 1977). This can be done globally to obtain $\langle K_z \rangle$ as defined by (13), or hemispherically, as defined by (15).

Globally, by combining (9), (10*a*) and (13),

$$-\langle K_z \rangle \frac{\partial \langle \chi \rangle}{\partial z} = \frac{1}{2} \int_{-\frac{1}{2}\pi}^{\frac{1}{2}\pi} \left(\overline{w\chi} - K_{yz} \frac{\partial \bar{\chi}}{\partial y} - K_{zz} \frac{\partial \bar{\chi}}{\partial z} \right) \cos\phi \, d\phi = \left\{ \langle \overline{w\chi} \rangle - \left\langle K_{yz} \frac{\partial \bar{\chi}}{\partial y} \right\rangle - \left\langle K_{zz} \frac{\partial \bar{\chi}}{\partial z} \right\rangle \right\}. \quad (23)$$

Then,

$$\langle K_z \rangle \frac{\partial \langle \chi \rangle}{\partial z} = A + \left\langle K_{zz} \frac{\partial \bar{\chi}}{\partial z} \right\rangle. \quad (24)$$

Since in (24) A is dependent on $\bar{\chi}$ and its horizontal but not its vertical gradient, species with these parameters equal must have different values of $\langle K_z \rangle$, which may be negative.

By using (9), (10*b*) and (15) a similar relation may be deduced for the hemispheric case:

$$\langle\langle K_z \rangle\rangle \frac{\partial \langle\langle \chi \rangle\rangle}{\partial z} = \left\langle\left\langle K_{yz} \frac{\partial \bar{\chi}}{\partial y} \right\rangle\right\rangle + \left\langle\left\langle K_{zz} \frac{\partial \bar{\chi}}{\partial z} \right\rangle\right\rangle - \langle\langle w \rangle\rangle \langle\langle \chi \rangle\rangle - \langle\langle \bar{w}^* \bar{\chi}^* \rangle\rangle, \quad (25)$$

where $\bar{w} = \langle\langle w \rangle\rangle + \bar{w}^*$ and $\bar{\chi} = \langle\langle \chi \rangle\rangle + \bar{\chi}^*$.

From (25) it follows that even when the hemispheric mean vertical motion is resolved into mean and eddy components, species with the same values and horizontal gradients (which are in general small compared to vertical gradients) but different vertical gradients will have different

$\langle\langle K_z \rangle\rangle$ values. If the species further have vertical gradients of opposite sign, the $\langle\langle K_z \rangle\rangle$ values may be of opposite sign.

In summary, the numerical work done with the use of (1)–(25) and the Meteorological Office two-dimensional (Comesa 1975; Murgatroyd 1977) and three-dimensional (Comesa 1975; O'Neill & Newson 1977) models was as follows:

(i) Mean Northern Hemispheric values of $\langle\langle w \rangle\rangle$ and $\langle\langle K_z \rangle\rangle$ (for O_3, H_2O and NO_x) were obtained by averaging three-dimensional model data at 6 h intervals over 30-day periods for a year (days 90–450 of the integration); the value for the first day of each month was also calculated.

(ii) The values of $\langle\langle w \rangle\rangle$ and $\langle\langle K_z \rangle\rangle$ for the first day of each month were similarly obtained for the Southern Hemisphere.

(iii) The zonal means $\bar{P}_{\text{true}}$ and $\bar{P}_{2D}$ defined by (21) and (22) were evaluated on days 180 and 360 (January and July solstices) of the three-dimensional integration.

(iv) The true column means $\langle k[NO][O_3] \rangle$ and $\langle\langle k[NO][O_3] \rangle\rangle$ in one dimension were also obtained on these days of the three-dimensional integration for the global, Northern and Southern Hemisphere cases, and compared with the one-dimensional model means $\langle k \rangle \langle [NO] \rangle \langle [O_3] \rangle$ and $\langle\langle k \rangle\rangle \langle\langle [NO] \rangle\rangle \langle\langle [O_3] \rangle\rangle$.

(v) The two-dimensional model data were averaged to produce one-dimensional data according to equations (23) and (25), for days 7380 (July solstice) and 7560 (January solstice) of an integration, at noon (the model included diurnal variations).

(vi) The same two-dimensional data were used to evaluate and compare the true global column mean $\langle \bar{k}\overline{[A]}\,\overline{[B]} \rangle$ of each of the rates of the reactions in table 1 with the one-dimensional model mean $\langle \bar{k} \rangle \langle \overline{[A]} \rangle \langle \overline{[B]} \rangle$. A similar calculation was done for the corresponding hemispheric means. It may be noted that the value of $\langle\langle K_z \rangle\rangle$ from the three-dimensional model was calculated as the hemispheric mean of negative values of the ratio of the local vertical eddy flux to the local vertical gradient of the appropriate tracer.

3. Results

A comparison of figures 1*a* and 1*b* shows the difference between monthly values obtained by averaging over 6 h intervals, and the values from a single day at the start of each month. The smoothing effect is not large; the data in figure 1*b* and *c* show that just 12 single-day values at monthly intervals draw up to a smooth contour diagram. Comparison of figures 1*b* and 1*c* shows that at any given time the Northern Hemisphere $\langle\langle w \rangle\rangle$ is closely balanced by a Southern Hemisphere $\langle\langle w \rangle\rangle$ of the same magnitude but opposite sign. A clear asymmetry between the hemispheres is revealed; the Northern Hemisphere summer has larger and more persistent positive $\langle\langle w \rangle\rangle$ values than the Southern Hemisphere summer, presumably because of the relatively greater extents of land in the Northern Hemisphere and of ocean in the Southern Hemisphere.

Figures 2*a* and 2*b* show the smoothing effect of the time averaging on $\langle\langle K_z \rangle\rangle$ (O_3) in the same way as did figure 1*a* and *b* for $\langle\langle w \rangle\rangle$. The differences are not large, and the general pattern in the Northern Hemisphere is of larger stratospheric values from November to April. Owing to the large computational burden, the Southern Hemisphere data have not been time-averaged, but if the same correspondence between the 12 single-day monthly values and the fully time-meaned values exists, figure 2*c* reveals a rather different structure for the mean hemispheric vertical eddy transport. Smaller values are widespread, and there is only a suggestion of the winter structure seen in figure 2*b*. These systematic differences between the hemispheres and seasons were noted in previous work (Tuck 1977) where (13) was applied to the separate hemispheres.

The time-meaned pattern of $\langle\langle K_z \rangle\rangle$ values occurring for NO_x and H_2O in the Northern Hemisphere are broadly similar to that for O_3, as may be seen by comparing figures 2*a*, 3*a* and 3*b*, but with some differences near the tropopause and in the upper stratosphere.

The Southern Hemisphere spot values for NO_x and H_2O corresponding to figure 2*b* and 2*c* show similar large areas of small values, with negative areas constituting a substantial fraction of the total.

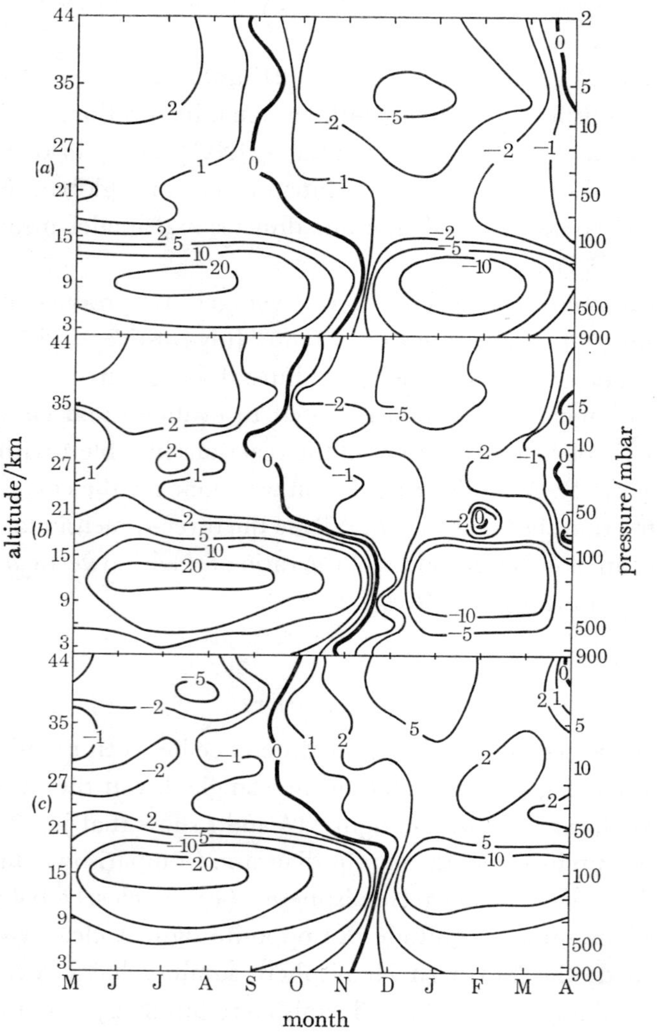

FIGURE 1. Mean hemispheric vertical velocities, three-dimensional model. Contours are 10^{-4} ms^{-1}. (*a*) N. Hemisphere, 30 day means from data at 6 h intervals; (*b*) N. Hemisphere, 30 day spot values; (*c*) S. Hemisphere, 30 day spot values.

A comparison of the hemispheric mean vertical velocities at winter and summer solstices reveals that in the troposphere the three-dimensional model values are considerably larger than the two-dimensional model values, as seen in figure 4. In the July stratospheres the values are quite similar, but in January the three-dimensionally derived values of $\langle\langle w \rangle\rangle$ are consistently larger than those obtained from the two-dimensional model data.

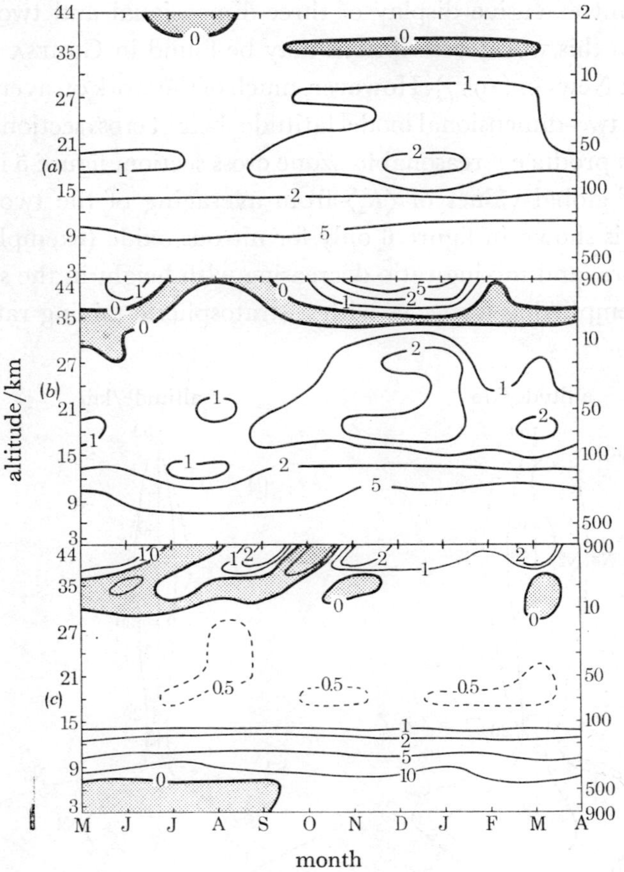

FIGURE 2. $\langle\langle K_z \rangle\rangle$ (O_3), three-dimensional model. Contours are $m^2 s^{-1}$; shaded areas are negative. (*a*) N. Hemisphere, 30 day means from data at 6 h intervals; (*b*) N. Hemisphere, 30 day spot values; (*c*) S. Hemisphere, 30 day spot values.

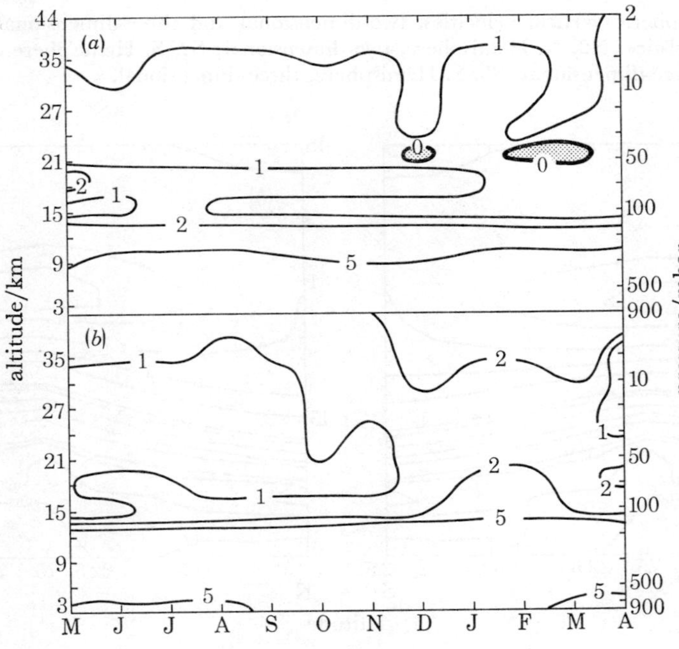

FIGURE 3. $\langle\langle K_z \rangle\rangle$ (H_2O) and $\langle\langle K_z \rangle\rangle$ (NO_x), three-dimensional model. Contours are $m^2 s^{-1}$. (*a*) $\langle\langle K_z \rangle\rangle$ (H_2O), 30 day means from data at 6 h intervals; (*b*) $\langle\langle K_z \rangle\rangle$ (NO_x), 30 day means from data at 6 h intervals.

Space does not permit extensive display of three-dimensional and two-dimensional model results used as inputs in this work; descriptions may be found in Comesa (1975), Murgatroyd (1977) and in O'Neill & Newson (1977). However, much of the work on average rates of chemical reactions was done with two-dimensional model latitude–height cross sections, and as an indicator of the model's ability to produce a reasonable ozone cross section, figure 5 is included.

The computation of global values of $\langle K_z \rangle$ from averaging of the two-dimensional model dynamical parameters is shown in figure 6 only for nitrous oxide (exemplifying a species well mixed in the troposphere, and mixing ratio decreasing with height in the stratosphere) and for total odd nitrogen (exemplifying a species with a stratospheric mixing ratio maximum and a

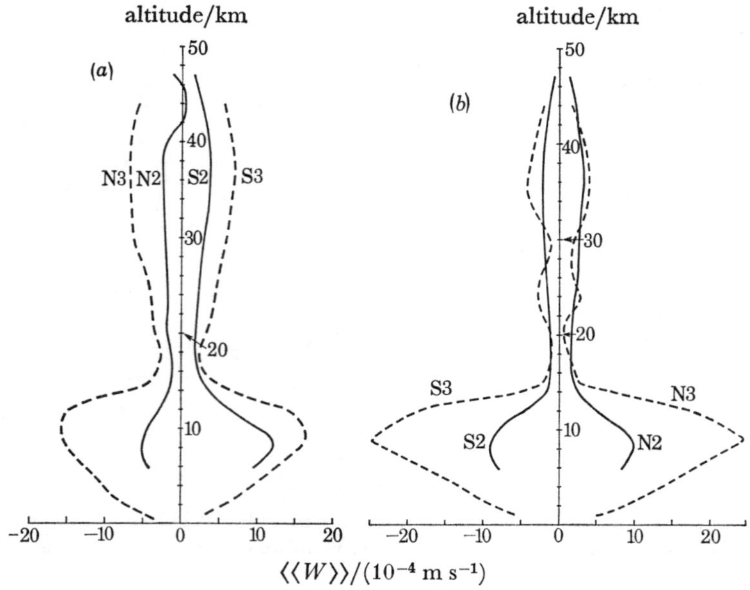

Figure 4. Mean hemispheric vertical velocities, two-dimensional and three-dimensional models. (a) January solstice; (b) July solstice. N2, N. Hemisphere, two-dimensional; S2, S. Hemisphere, two-dimensional; N3, N. Hemisphere, three-dimensional; S3, S. Hemisphere, three-dimensional.

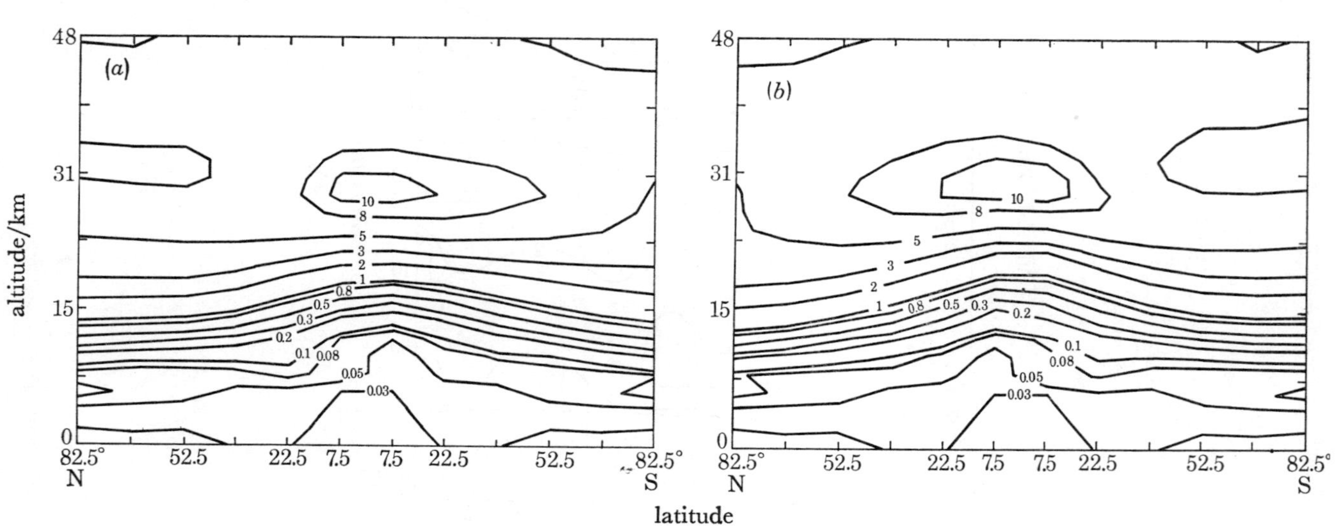

Figure 5. Two-dimensional model (Z, ϕ) ozone cross sections. (a) O_3 volume mixing ratio contours, solstice, year 21; (b) O_3 volume mixing ratio contours, July solstice, year 20.

downwards directed gradient above the tropopause). The profiles are different for the two types of tracer, and $\langle K_z \rangle$ is not well behaved near the tropopause for N_2O, or near the photochemical source region for $(NO+NO_2+HNO_3)$. Similar results were obtained for H_2O (like N_2O) and O_3 (like odd nitrogen, but more variable). An attempt to obtain hemispheric mean values for a total flux-based $\langle\langle K_z \rangle\rangle$ from (13) produced profiles with large negative excursions and generally erratic behaviour.

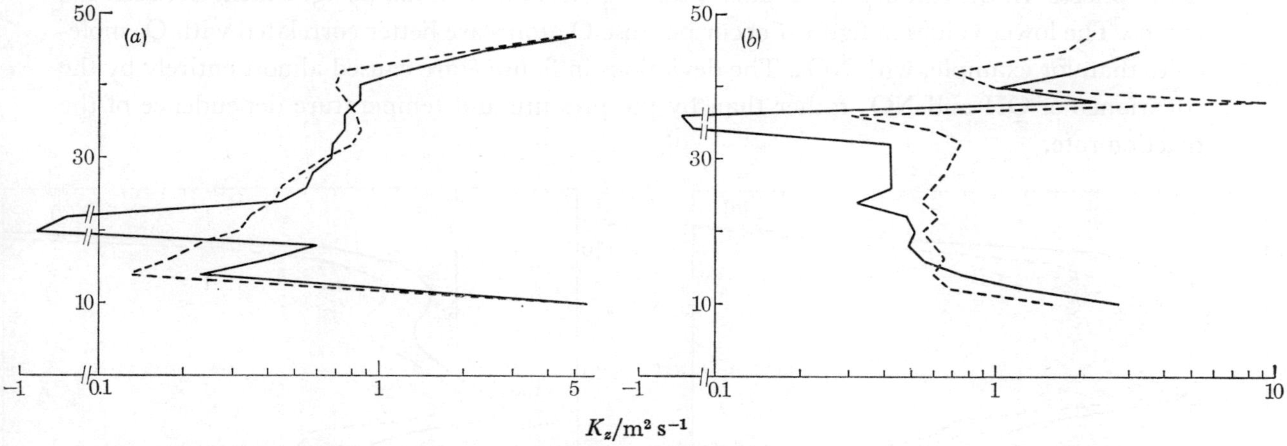

FIGURE 6. K_z profiles from averaging two-dimensional model data. (*a*) $\langle K_z \rangle$ (N_2O), January and July solstices; (*b*) $\langle K_z \rangle$ $(NO+NO_2+HNO_3)$, January and July solstices. Solid line, January; broken line, July.

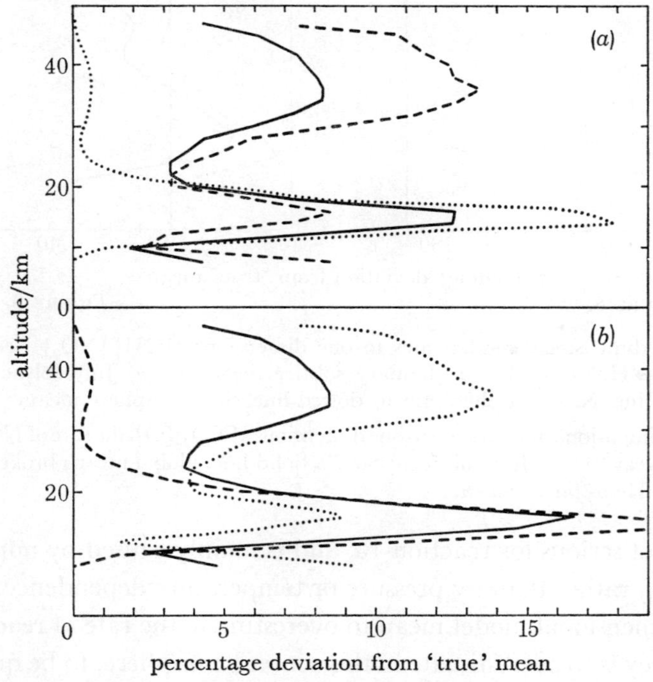

FIGURE 7. Averaging two-dimensional model rates to one dimension;

$$k[O]\,[O_2]\,[M], \quad \text{the rate of} \quad O+O_2+M \longrightarrow O_3+M.$$

(*a*) January solstice, year 21; (*b*) July solstice, year 20. Solid line, global mean; broken line, N. Hemisphere mean; dotted line, S. Hemisphere mean.

Although the averaging of two-dimensional model fields of molecular number densities, rate coefficients and reaction rates was carried out for all the reactions in table 1 (except photo-dissociations), the results are illustrated here only for reactions 8, 11 and 13; figures 7, 8 and 9 respectively show the percentage deviation of the one-dimensional model mean $\langle \bar{k} \rangle \langle \overline{[A]} \rangle \langle \overline{[B]} \rangle$ from the 'true' global mean $\langle \bar{k}\overline{[A]}\,\overline{[B]} \rangle$. A consistent pattern is clear: there are at both times, in global and hemispheric cases, significant deviations in the upper troposphere and lower stratosphere. In the winter hemisphere, there are significant deviations in the middle and upper stratospheres. In the summer hemisphere, this region shows reasonable agreement between the means. The lower values in figure 7 occur because O atoms are better correlated with O_3 molecules than for example, with NO_2. The deviations in figure 8 are caused almost entirely by the covariance of OH and NO_2 rather than by the pressure and temperature dependence of the reaction rate.

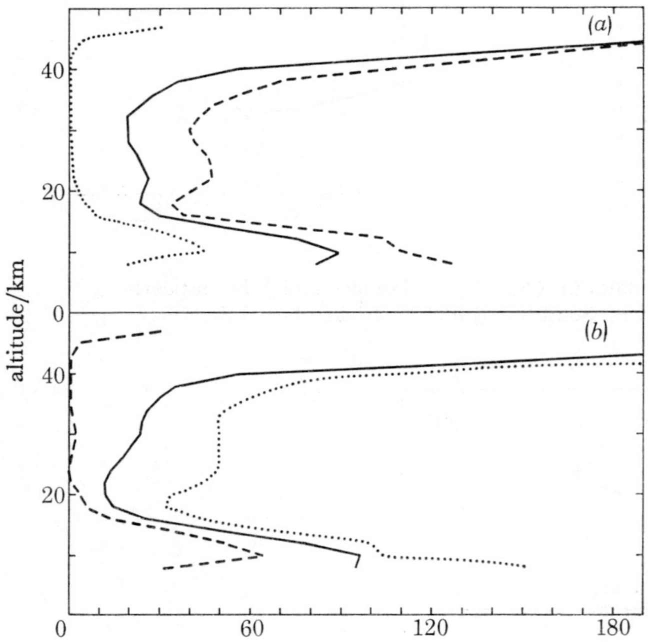

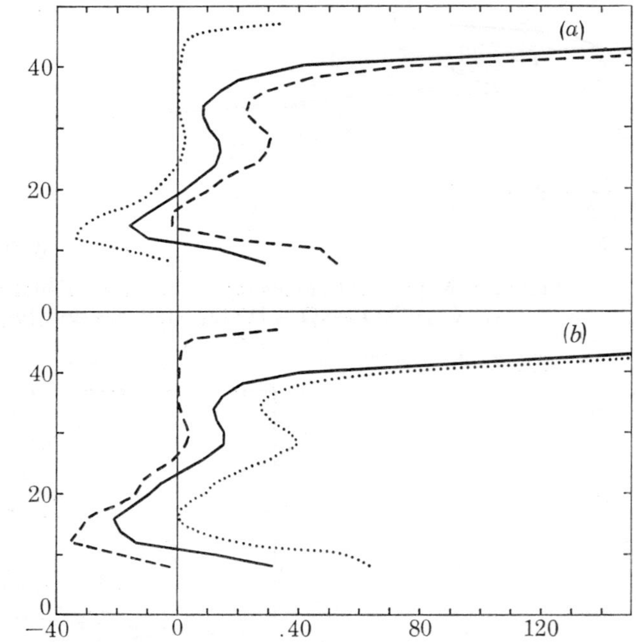

percentage deviation from 'true' mean

FIGURE 8. FIGURE 9.

FIGURE 8. Averaging two-dimensional model rates to one dimension; $k[OH]\,[NO_2]$ (see table 1), the rate of $OH + NO_2(+M) \longrightarrow HNO_3(+M)$. (*a*) January solstice, year 21; (*b*) July solstice, year 20. Solid line, global mean; broken line, N. Hemisphere mean; dotted line, S. Hemisphere mean.

FIGURE 9. Averaging two-dimensional model rates to one dimension; $k[NO_2][O]$, the rate of $NO_2 + O \longrightarrow NO + O_2$. (*a*) January solstice, year 21; (*b*) July solstice, year 20. Solid line, global mean; broken line, N. Hemisphere mean; dotted line, S. Hemisphere mean.

The deviation is most serious for reaction 13, and is mainly caused by nonlinear correlations between OH and NO_2, rather than by pressure or temperature dependence. The nature of the effect is for the one-dimensional model mean to overestimate the rate of reactions 8 and 13; for reaction 11 the tendency is to overestimate in the winter hemisphere, to be quite accurate in the summer stratosphere, and to underestimate in the summer tropopause region.

The quantity $100\,(\bar{P}_{true} - \bar{P}_{2D})/\bar{P}_{true}$, the percentage difference between the true and two-dimensional model means, was evaluated at January solstice (360 days) and July solstice (180 days) of the three-dimensional model integration. The resulting data are plotted respectively in figures 10*a* and 10*b*, and clearly display the regions of inadequacy of two-dimensional models in calculating the rate of reaction of aircraft injected NO_x with O_3. In agreement with the two-dimensional model averaged to one dimension, the only region of accuracy is the middle stratosphere in summer. It is of interest that in January the three major regions of negative anomaly in the zonal mean reaction rate are associated with a westerly jet; the region of positive anomaly is in the broad location of the source of aircraft emission of NO_x. Virtually all the deviation in $\bar{P}_{2D}$ arises from nonlinear correlation between NO and O_3, with only a small component from the products k[NO] and $k[O_3]$. The July zonal means of the reaction rate again show large negative anomalies associated with the tropospheric jets; there is, however, no area of negatives associated with the Southern Hemisphere polar night vortex. A very intense area of large positive anomalies (up to 80 %) occurs near the surface in mid-latitudes, particularly in the Northern Hemisphere. The covariance $\overline{[H_2O]'[O_3]'}$ caused a different effect; the quantity $100(\bar{P}_{true} - \bar{P}_{2D})/\bar{P}_{true}$ had large positive values up to 90 % around tropospheric westerly jets, and negative values as low as -20 % above the equatorial tropopause.

TABLE 2. THREE-DIMENSIONAL DATA AVERAGED TO ONE DIMENSION

$\langle k\rangle\langle[NO]\rangle\langle[O_3]\rangle/\langle k[NO][O_3]\rangle$ and $\langle\langle k\rangle\rangle\langle\langle[NO]\rangle\rangle\langle\langle[O_3]\rangle\rangle/\langle\langle k[NO][O_3]\rangle\rangle$.

Single angle brackets represent a global average, and double angle brackets a hemisphere average.

global Jan.	N. Hem. Jan.	S. Hem. Jan.	global July	N. Hem. July	S. Hem. July	Pressure mbar†
0.985	0.891	1.039	0.822	0.993	0.685	2
1.057	0.912	1.135	0.762	1.022	0.647	5
1.122	1.024	1.191	0.869	1.036	0.774	10
1.311	1.232	1.440	1.041	1.106	1.064	20
1.264	1.181	1.426	1.066	1.089	1.230	30
1.235	1.187	1.461	1.091	1.088	1.601	50
0.825	0.795	1.273	0.687	0.710	1.528	70
0.508	0.533	1.053	0.471	0.497	1.378	100
0.397	0.524	1.116	0.456	0.458	1.418	200
0.566	0.659	1.379	0.687	0.583	1.404	300
0.885	0.912	1.392	1.205	0.881	1.516	500
0.938	0.966	1.459	1.315	0.936	1.430	700
0.992	1.016	1.466	1.942	1.147	1.439	900

† 1 mbar = 10^2 Pa.

Figure 10*a* and 10*b* demonstrate that two-dimensional models will not be able to model tropospheric chemistry in a quantitative manner; problems also occur in the lower stratosphere, particularly in winter.

The last results are shown in table 2, and consist of the three-dimensional data for $k[NO][O_3]$ averaged down to one dimension. The data clearly show hemispheric asymmetries, with ± 50 % deviations from the true mean occurring. The meridional variance has in some cases, particularly in the Southern Hemisphere, outweighed the zonal variance causing the anomalies in figure 10*a* and 10*b*. The data of table 2 may be compared with figures 7–9, which display the equivalent one-dimensional results from averaging the two-dimensional model.

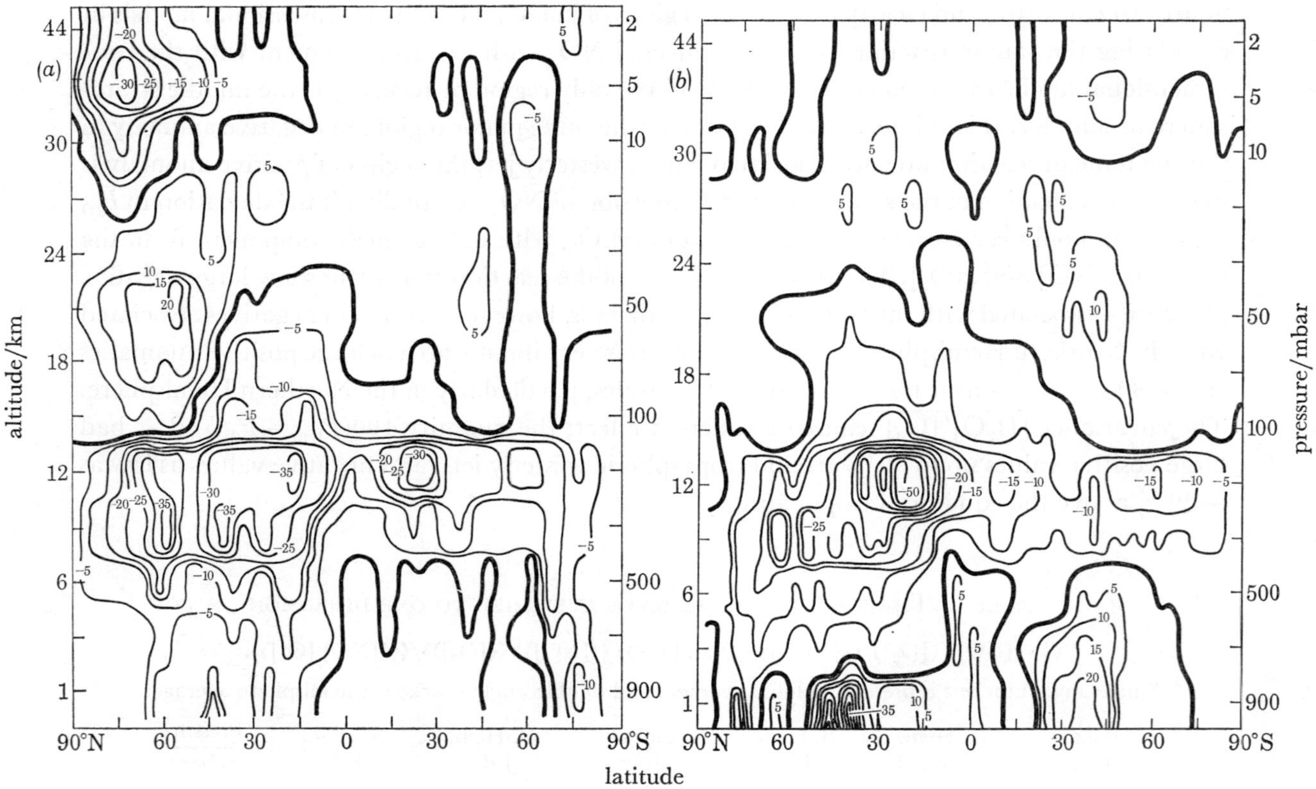

FIGURE 10. Averaging three-dimensional model rates to two dimensions;

$k[NO][O_3]$, the rate of $NO + O_3 \longrightarrow NO_2 + O_2$.

Contours are percentage deviation of $\overline{k}\,\overline{[NO]}\,\overline{[O_3]}$ from $\overline{k[NO][O_3]}$. [NO] is effectively a gaseous tracer for aircraft emissions in the three-dimensional model. (*a*) January solstice, day 360; (*b*) July solstice, day 180.

4. Discussion and conclusions

Before considering in detail the results obtained in § 3, it is essential to realize that the ideal way of computing the zonal, hemispheric or global mean of a particular quantity is by using local measurements distributed throughout the real atmosphere. Should the very large volume of data needed for this task ever be acquired, it would then be possible to check experimentally the averaging assumptions incorporated in models of varying dimensionality. Meanwhile, it can be said that as regards chemical composition, the real atmosphere is likely to be inhomogeneous and to have nonlinear correlations between species on a much larger range of scales than is encompassed by the models used here. This is particularly true of the smaller scales, which tend to be missing or badly represented in the models. The estimates of the deviations of zonal, hemispheric and global mean rates of reactions from their 'true' values obtained in this work are therefore likely to be lower than is really the case in the atmosphere. To this extent, they therefore represent an optimistic picture of the ability of current models to compute mean rates of chemical reactions in the atmosphere, and perturbations caused by gases emitted as a result of human activities.

Considering now the specific results described in § 3, it is clear that as far as the three-dimensional model is concerned, there is a distinct asymmetry between the hemispheres in the atmospheric dynamics when these are averaged down to a single column (representing a hemispheric mean). The upward vertical velocities occurring in the Northern Hemisphere summer troposphere have larger magnitudes than those occurring in the Southern Hemisphere summer troposphere. It is apparent from figure 1 that mean upward velocities in the Northern Hemisphere model troposphere (balanced by downward velocities in the Southern Hemisphere troposphere) persist for about $7\frac{1}{2}$ months from May to November inclusive; there are thus approximately $4\frac{1}{2}$ months in the Northern Hemisphere with mean tropospheric vertical velocities direct downwards. In the stratosphere, the position is different. The larger values tend to occur in the Southern Hemisphere summer, and persist in an upwards direction for 6–7 months from October to April, with corresponding negative values in the Northern Hemisphere winter. The results must be viewed with caution, since the Southern Hemisphere simulation in the troposphere is not as good as the Northern Hemisphere simulation (COMESA 1975). The subtropical jet was too far north, and tropospheric eddy activity (cyclones and anticyclones) was concentrated not at about 60 °S, but was also too far north. The three-dimensional model integration of 450 days contained two northern winters but only one southern one, and it is possible that a second southern hemisphere winter might have been an improvement. However, both Northern Hemisphere winters included periods where sudden warming events occurred, whereas none occurred in the Southern Hemisphere. Both Northern and Southern Hemisphere stratospheres were reasonable simulations, with some shortcomings such as too strong a polar night vortex.

It is perhaps worth stressing that the fundamental hemispheric asymmetry in the model derives from the distribution of land and oceans, and the consequent differences in surface temperature and exchanges. In the real atmosphere, effects such as aircraft operations and chlorofluoromethane releases are heavily biased towards the Northern Hemisphere; thus, subject to the admitted imperfections of the three-dimensional model, it may be concluded that hemispheric asymmetries should be taken into account in perturbation calculations. This is highlighted by the fact that values of the hemispheric mean vertical velocity, $\langle\langle w\rangle\rangle$, are heavily influenced (because of $\cos\phi$ weighting) by the upward branch of the Hadley Cell occurring in tropical regions, and the upward extension of this motion is in turn the largest component in the air inflow to a hemispheric stratosphere (Reiter 1975). Manabe & Mahlman (1976) have also noted substantial asymmetries between hemispheres in a general circulation model integration.

The Northern Hemisphere pattern of higher values of up to $2\ m^2\ s^{-1}$ during winter (when there was a sudden warming) and values from 0.2 to $1.0\ m^2\ s^{-1}$ in summer shows a clear seasonal variation in the eddy transfer, for all three tracers. The rather different vertical eddy transfer coefficients, K_z, derived for the Southern Hemisphere are of interest. It has been noted previously (Tuck 1977) that there is no *a priori* reason to expect K-theory to apply to vertical transfer in the atmosphere, and hence negative values are not excluded in principle and in fact occur.

The vertical eddy transfer is then seasonally dependent in the Northern Hemisphere, as are the vertical motions. Thus, on a hemispheric basis, faster chemistry in the summer is correlated with mean upward motion and slower vertical eddy transfer. The reverse situation pertains in winter with slower photochemistry, mean downward motion and faster eddy transfer. No one-dimensional model to date has accommodated these correlations.

The eddy transfer coefficients, K_z, from the three-dimensional model show a sensitivity to the location of the tracer sources and sinks. Mahlman (1976) has reached a similar conclusion. The values of K_z derived from averaging the two-dimensional model data and dynamical parameters show clear disruption near the tropopause for nitrous oxide, and immediately above the ozone layer maximum for odd nitrogen. The profiles are different for different tracers, and different at summer and winter solstices for the same tracer. These conclusions are again serious for one-dimensional modelling, since models to date have used a single, time-invariant K_z profile.

The results of averaging two-dimensional model fields to obtain hemispherically and globally averaged rates of reaction showed systematic deviations from the 'true' global mean $\overline{k[A]\,[B]}$ by the one-dimensional model global mean $\bar{k}\overline{[A]}\,\overline{[B]}$. These deviations were larger in the winter hemisphere, and generally showed a maximum near the tropopause. Within the model limitation, it may be concluded from figures 7–9 that a one-dimensional model only computes the rate of a gas reaction accurately in the middle and upper stratosphere in summer. This is when the photochemistry is dominant, and it may be concluded that at other times and in other places the two-dimensional model dynamics are giving rise to nonlinear correlations between species which cause serious deviations in the one-dimensional average rate.

The conceptual errors occurring here in the estimation of the rates of individual gas kinetic processes will, in a model calculation, combine to produce a cumulative effect. The magnitude of the error so induced in any particular photochemical calculation could only be assessed accurately for a one-dimensional model by running it in models of two and three dimensions. Inspection of figures 7–10 and table 2 suggests that it could be quite large.

The association of the maximum negative anomalies in the zonal mean rate $\bar{k}\overline{[NO]}\,\overline{[O_3]}$ (calculated from the three-dimensional model), with westerly jet streams in figure 10*a*, strongly suggests that the nonlinear correlations between $[NO_x]$ and $[O_3]$ are caused by large-scale eddy fluxes, which also tend to be highly correlated with jet maxima. The consequences for atmospheric chemistry are considerable. For example, the rate of any reaction in the upper troposphere or winter stratosphere cannot be reliably computed in even a two-dimensional model; the concept of a single mean tropospheric lifetime for a species reacting in a temperature-dependent manner with a tropospheric mean value of [OH], for example, is seriously undermined. Nonlinear correlations near corridors of aircraft operation in the lower stratosphere render two-dimensional model calculations inaccurate in computing the rate at which aircraft-emitted nitrogen oxides will react with ozone. The further averaging to one dimension makes one-dimensional averages worse still, with the meridional variance sometimes opposing the zonal variance, and at other times reinforcing it.

It seems likely that while eddy vertical transfer may be significant (but not dominant) in the three-dimensional model Northern Hemisphere, vertical transfer of materials in the Southern Hemisphere is achieved to a greater extent by the mean motion; other significant asymmetries involve the extent and persistence of seasonal variations in mean hemispheric vertical velocities.

The deviations occurring from 'true' mean rates of reaction in the course of averaging both three- and two-dimensional models suggest a consistent picture of maximum anomalies occurring in regions where dynamical features tend to be prominent: for example near jet streams, the tropopause region and in the winter stratosphere.

A factor which may explain the large errors in zonal means near tropospheric jet streams is the structure of tropospheric depressions in mid and high latitudes. These phenomena are regarded as eddies on the global scale, but on a scale of *ca.* 10^3 km possess distinct organization, with a warm sector from low latitudes separated from colder high latitude air by frontal systems, in which rainout of soluble species may occur (the three-dimensional model NO_x had a rainout sink). Tropopause folding and stratosphere–troposphere interchanges are known to be associated with depressions, fronts and jet streams; since warm sector air probably has a different chemical composition from cold sector air, nonlinear correlations may occur in these localities. The deviations near the polar night jet in the Northern Hemisphere may occur because the sudden warming similarly exchanges large volumes of air of different origins (and therefore different chemical compositions). The errors near the Southern Hemisphere polar night vortex are much smaller, possibly because no sudden warming event occurred.

It should be realized that the real atmosphere is likely to be more inhomogeneous than even the three-dimensional model. For example, according to Riehl & Malkus (1958) the upward branch of the Hadley circulation may be accounted for globally by the operation of a few thousand cumulonimbus cores in the tropics. These events, occurring on a scale too small for explicit representation in a three-dimensional g.c.m., thus collectively form the dynamical feature which is the major source of air for the stratosphere (Reiter 1975). It may be true that they are sources of nitric oxide, produced by lightning discharges (Tuck 1976); they are certainly a sink for soluble tropospheric gases by means of rainout processes, see, for example, Junge (1963). There are thus likely to be significant departures from homogeneity as regards chemical composition in these events which are significant globally but too localized for specific formulation in a g.c.m.

Burnett (1976) has recorded observations of OH column densities fluctuating by factors between 5 and 10 during daylight hours, from day to day. Such observations are difficult to explain in any way other than by highly nonlinear correlations between the number densities of different gases. It follows that it is of great interest to obtain simultaneous measurement of molecular concentrations in both time and space; simultaneous observations of NO_2, O_3 and OH column densities from the ground are currently feasible, for example.

Since the checking of model calculations of rates of reaction by measurements is likely to be limited for some time to come, it may be worth examining an alternative modelling approach. It should be possible to compute Lagrangian air trajectories from meteorological observations, and perform model chemical calculations as a function of latitude, longitude, pressure and temperature along the path of the air parcel chosen, including if necessary some representation of the effect of any cloud or precipitation processes observed. Comparison of different trajectories would then yield information about correlations between species.

The final observation which may be made as a result of this examination of model assumptions and data is that one and two-dimensional models of the sort widely used for perturbation calculations of the effect of aircraft and chlorofluoromethanes on the stratospheric ozone balance (C.I.A.P. 1974; Booker 1975; Comesa 1975; N.R.C. 1976) are unreliable tools for quantitative estimation of rates of gas reactions in the atmosphere. This is because nonlinear correlations, arising largely from dynamical phenomena, are either ignored or inadequately represented. The resultant errors come from the design of one- and two-dimensional models rather than from the input data, and so have not been included in published estimations of error limits associated with ozone perturbation calculations.

The author wishes to express his appreciation to R. J. Murgatroyd for many valuable discussions and suggestions. Thanks are also due to M. K. Hinds and S. A. Clough for two-dimensional model data in advance of publication. D. W. Jones, K. S. Groves and D. Jerrett all helped with the computation at various times.

References (Tuck)

Burnett, C. R. 1976 *Geophys. Res. Lett.* **3**, 319–322.

Booker, H. G. 1975 *Environmental impact of stratospheric flight.* Washington, DC: Climatic Impact Committee, National Academy of Sciences.

C.I.A.P. 1974 *Climatic Impact Assessment Program. Report of findings.* Cambridge, Mass.: U.S. Department of Transportation.

Comesa 1975 *The report of the Committee on Meteorological Effects of Stratospheric Aircraft 1972–1975.* Bracknell: U.K. Meteorological Office.

Gudiksen, P. H., Fairhall, A. W. & Reed, R. J. 1968 *J. geophys. Res.* **73**, 4461–4473.

Junge, C. E. 1963 *Air chemistry and radioactivity.* New York: Academic Press.

Luther, F. M. 1973 *A.I.A.A./A.M.S. Int. Conf., Denver, Colorado*, A.I.A.A. paper 73–498.

Mahlman, J. D. 1976 *Proc. 4th Conf. on C.I.A.P.*, DOT-TSC-OST-75-38. Cambridge, Mass.: U.S. Department of Transportation.

Manabe, S. & Mahlman, J. D. 1976 *J. Atmos. Sci.* **33**, 2185–2217.

Murgatroyd, R. J. 1977 *Proc. Scient. Seminar on Stratospheric Monitoring.* Paris Sécretariat d'Etat Auprès du Ministre de l'Equipement (Transports).

N.R.C. 1976 *Halocarbons: effects on stratospheric ozone.* Washington, D.C.: National Academy of Sciences.

Newell, R. E., Kidson, J. W., Vincent, D. G. & Boer, G. J. 1974 *The general circulation of the tropical atmosphere*, vols 1 and 2. Cambridge, Mass.: M.I.T. Press.

O'Neill, A. & Newson, R. L. 1977 *IAMAP, Comm. Met. Upper Atmos., Coll. Ext. Summ.*, Seattle, pp. 50-1–50-6.

Reed, R. J. & German, K. E. 1965 *Mon. Weath. Rev.* **93**, 313–321.

Reiter, E. R. 1975 *Rev. Geophys. Space Phys.* **13**, 459–474.

Riehl, H. & Malkus, J. S. 1958 *Geophysica* **6**, 505–538.

Tuck, A. F. 1976 *Q. Jl R. met. Soc.* **102**, 749–756.

Tuck, A. F. 1977 *Proc. R. Soc. Lond.* A **355**, 267–299.

Discussion

R. A. Scriven (*Central Electricity Research Laboratories, Kelvin Avenue, Leatherhead, Surrey*). Dr Tuck has demonstrated very clearly that some care is required in constructing global averages and interpreting the results from sample models. Does this not mean that the measurement of a few variables at a few points on the Earth cannot be taken to describe events over the Earth as a whole? Is the information on CO_2, for example, sufficient for us to be able to say with certainty that the total amount of CO_2 in the Earth's atmosphere is increasing at 0.25 % per annum? Perhaps Dr Tuck's models can be used to tell us what the existing measurements really do mean and what further measurements are required.

A. F. Tuck. The covariance of reactants will mean that it may be difficult to estimate a global budget for reactive species with short lifetimes using measurements at a few locations. For species with longer lifetimes, the effectiveness will depend on the patchiness of the distribution of its sources and sinks. In the case of CO_2, it is my impression that a small number of stations would give a reliable estimation of global amounts.

D. T. Swift-Hook (*Central Electricity Research Laboratories, Kelvin Avenue, Leatherhead, Surrey*). Dr Tuck's diagrams showed that vertical variations of the different quantities in time and place are apparently striated in about half a dozen steps. Have these layers any physical reality or are they simply artefacts of the computing, due perhaps to a coarse mesh?

A. F. Tuck. It must be remembered that the three-dimensional model has 13 levels up to 44 km; the vertical resolution of about 3 km is rather coarse for really good reproduction of features like jet streams and the tropopause.

Phil. Trans. R. Soc. Lond. A. **290**, 495–504 (1979)
Printed in Great Britain

The cycle of atmospheric nitrous oxide

By J. Hahn
Max-Planck-Institut für Chemie (Otto-Hahn-Institut), Mainz, Germany

An attempt has been made to construct an N_2O balance of the troposphere in order to demonstrate the state of knowledge of the cycle of atmospheric N_2O. This balance has been constructed considering the parameters relevant for atmospheric N_2O as far as known such as the horizontal and vertical distribution in the atmosphere and in the oceans, the variation of the atmospheric mixing ratio with time, the fluxes in the air–sea and in the air–soil systems, and man-made fluxes. The total atmospheric burden of N_2O appears to be in the range 1600–2100 Mt, the tropospheric turn-over time between 4 and 12 years.

Most of the atmospheric N_2O seems to originate from microbial activities at the Earth's surface with the oceans apparently being the major source with about 35 % of the total source as obtained from the tropospheric N_2O mass and turn-over time (range of uncertainty 5–100 %) followed by soil and fresh water with about 25 % of the total source (range of uncertainty 4–100 %) including the contribution of industrial nitrogen fertilizers with about 10 % of the total source (range of uncertainty 2–24 %). Direct man-made N_2O sources seem to be relatively small with about 2 % of the total source (range of uncertainty 0.5–5 %), but may grow considerably in the future. The N_2O production by chemical reactions in the troposphere itself (e.g. electrical discharges) is not known; a figure of between 3 and 70 % of the total source has been inferred from laboratory experiments.

The sinks of atmospheric N_2O are poorly known. The only known major N_2O destruction appears to occur in the stratosphere. The rate of transport into the stratosphere, however, limits this sink to about 10 % of the total amount of N_2O to be destroyed under steady state conditions (range of uncertainty 3–22 %). The major problem of the cycle of atmospheric N_2O is the unknown main sink. Since the stratosphere does not seem to come into question, the missing sink should be found somewhere in the troposphere, pedosphere, and/or hydrosphere.

Introduction

The past decade has revealed that the stratospheric ozone layer is one of the most delicate features of the global environment. The ozone layer is important to life on Earth, because it absorbs potentially harmful solar radiation in the u.v. B range (280–320 nm) that otherwise would reach ground level. Small decreases in the thickness of this ozone shield may not only lead to increases in the incidence of human skin cancer, but may also have adverse effects on plant growth and animal life. Solar ultraviolet radiation can penetrate 10 m in clear water; and phytoplankton (the microscopic green plants which are the base of the aquatic food chain and one of the main sources of free oxygen) is sensitive to u.v. radiation. On the other hand, however, the latitudinal and temporal variability of the total vertical column density of ozone implies that there are rather effective biological repair mechanisms against permanent damage which may result from increased u.v. stress. It is difficult, if not impossible at present, to assess the ecological effects of long-lasting reductions in global ozone, nor can one predict the effects of a long-lasting global ozone reduction on climate (C.I.S.C. 1976).

Four different potential hazards to the ozone shield have been identified in recent years: nuclear weapon tests in the stratosphere, emissions from supersonic transports, upward diffusion of chlorofluoromethanes used mainly as aerosol propellants, and, last but not least, upward diffusion of nitrous oxide (N_2O). While the first three hazards are exclusively man-made, nitrous oxide is found to be a natural component of the Earth's atmosphere and appears to be responsible for the destruction of a major amount of the ozone formed in the stratosphere by photodissociation of molecular oxygen. The existing ozone layer is made of what is left over from ozone photolysis and reactions with OH and HO_2 radicals, with nitric oxide, and with natural halogen compounds such as methyl chloride. The nitric oxide mainly originates from N_2O (Crutzen 1970; Johnston 1971). N_2O may pose a threat to the existing ozone layer if its atmospheric level rises due to large-scale soil manipulations including development of virgin land and increasing fertilization with nitrogen-containing fertilizers. An increase in atmospheric N_2O might also be brought about by increasing industrial production of nitric acid from ammonia Hahn (1975) and by increasing combustion of fossil fuels (McElroy *et al.* 1977; Hahn & Junge 1977).

For this reason, N_2O has become a subject of major concern among atmospheric scientists. The discussion of the last two years has revealed how little we know about the cycle of atmospheric N_2O, despite the considerable effort made in the past (Goody 1969; Schütz *et al.* 1970; LaHue *et al.* 1970; Junge *et al.* 1971) and in recent years (see, for example, Murcray *et al.* 1973; Harries *et al.* 1974; Ehhalt *et al.* 1974; Hahn 1974, 1975; Yoshinari 1976; Rasmussen *et al.* 1976*a, b*; Schmeltekopf *et al.* 1977).

In this paper, an attempt has been made to summarize what is known about the processes which drive the cycle of atmospheric N_2O and about the fluxes involved.

Processes of N_2O formation and N_2O destruction

Figure 1 shows a summary of the aspects of the global nitrogen cycle which are relevant for atmospheric N_2O. In order to avoid overcrowding, only major features are included. Reactions which may occur in the troposphere involving the oxidation of ammonia and the eventual formation of nitric oxide (NO) and/or N_2O by lightning have been omitted.

The figure is made of a system of three interlinked concentric rings (8–7–9–10–11–10–9–7, 1–2–3–4–5–4–3–2, and 6) which represent biological and abiological processes. The wavy line separating the outer ring from the inner parts of the figure indicates the air–soil(–sea) boundary. Encircled by the wavy line are microbiological processes occurring in soil, fresh water, and sea water. The outer ring shows what happens to N_2O in the stratosphere.

The total cycle consists of several subcycles each of which has its own rate. Subcycle I appears to be the slowest and proceeds via 1–2–3–7*a*–7*b*–9–11–1. The cycling time is primarily determined by the rate of N_2O transport from the Earth's surface into the stratosphere (7*a*–7*b*) and the rate of the reverse transport of nitrate (11–1). Recent studies suggest an average cycling time on the order of 10 years. The same holds for the other branch of this subcycle (1–2–3–7*a*–7*b*–8*b*–8*a*–5–6–5–4–3–2–1) where again the transport in and out of the stratosphere (7*a*–7*b* and 8*b*–8*a*) should determine the cycling time.

Subcycle II proceeding via 1–2–3–7*a*–8*a*–5–6–5–4–3–2–1 appears to be faster. Here the cycling time is primarily determined by that part of subcycle II in which both higher plants and animals and various microorganisms participate (5–6–5). The average cycling time may be estimated to be on the order of months.

Subcycle III involving the steps 1–2–3–4–5–6–5–4–3–2–1 may be even faster than subcycle II, because it is shorter. A conservative estimate of the average cycling time would be in the order of weeks. Both subcycle II and subcycle III should be fastest when only microorganisms are involved. The steps 5–6–5 actually form another subcycle which although it often operates separately may be considered to be part of the three subcycles mentioned.

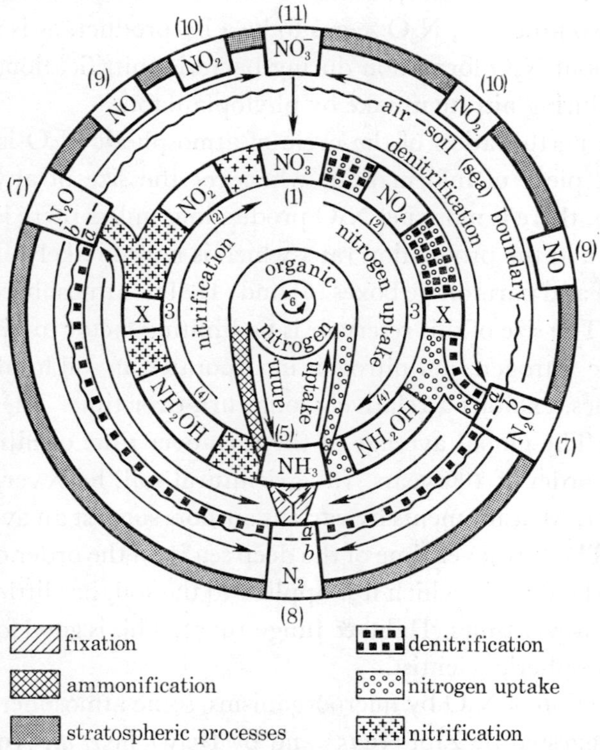

FIGURE 1. Global nitrogen cycle: features relevant for atmospheric N_2O. Wavy line separates processes in soil and in aquatic environments shown in the centre of the figure from atmospheric processes. Dashed line in the boxes of N_2O and N_2 symbolizes exchange of N_2O and N_2, respectively, across the air–soil(–sea) boundary layer. The reaction pathways of the various processes are marked by different symbols. (From Hahn & Junge 1977.)

N_2O is formed in step 3–7*a* and is either lost to the atmosphere (7*a*–7*b*) or reduced to molecular nitrogen (7*a*–8*a*) which is again either lost to the atmosphere (8*a*–8*b*) or reduced to ammonia (8*a*–5) by nitrogen fixation and returned to the cycles of the inner part of the figure.

In the stratosphere (outer ring of the figure), N_2O is destroyed by photolysis and reaction with atomic oxygen $O(^1D)$. The products of this reaction are molecular nitrogen (8*b*) and NO (9). NO reacts with ozone and ultimately forms nitrate (11) which is transported back into the troposphere where it is washed out by rain or returned to the Earth's surface by dry deposition (11–1).

Most of the ammonia nitrogen (5) goes through the loop 5–6–5 and is trapped in the pool of organic nitrogen compounds (6) for various lengths of time. The pool of organic nitrogen compounds contains living matter in the form of microorganisms, plants, and animals, and dead organic matter in the form of humus, various excreta, and particulate and dissolved organic matter in aquatic environments.

The microbiological processes relevant for the production of N_2O are symbolized by the steps 1–2–3–7*a*–8*a*, representing the process of bacterial denitrification, 5–4–3–(7*a*)–2–1, representing the process of bacterial nitrification, and 1–2–3–(7*a*)–4–5–6, representing nitrogen uptake from nitrates by phytoplankton. These three processes seem to produce most of the N_2O present in the atmosphere; bacterial denitrification is commonly believed to be the main source. In bacterial denitrification, N_2O may be taken to be an obligatory intermediate (Hahn & Junge 1977). Depending on the environment, changing amounts of N_2O and molecular nitrogen are formed. In the other two processes, N_2O seems to be a by-product, as is indicated in the figure. Very little is known about N_2O formation during bacterial nitrification, and virtually nothing about N_2O formation during nitrate uptake by phytoplankton.

As far as man-made perturbation of the cycle of atmospheric N_2O is concerned, the figure contains an important piece of information: whenever the size of any of the various boxes (reservoirs) is increased, there is more net N_2O production, unless there is feedback between the atmospheric N_2O level and the production rate, which is not too likely. This holds for all of the boxes except for box (8) and parts of the boxes (6) and (1). The atmosphere is the largest reservoir of molecular nitrogen. The size of this reservoir is no limiting factor to nitrogen fixation. Part of the reservoir of organic nitrogen, i.e. nitrogen incorporated in soil humic matter in temperate and boreal climate zones, exhibits a relatively long turn-over time. The same holds for nitrate found in the deep sea. The global average of the turn-over time of nitrogen in natural soil is estimated to be on the order of 400 years. In agricultural soil, however, the turn-over time of nitrogen is much shorter. Measurements by various authors suggest an average of about 50 years (McElroy *et al.* 1977). The turn-over time of the deep sea is on the order of 1000 years. Fertilizer nitrogen, regardless of the form in which it is applied to the soil, has little chance to get into the reservoirs with long turn-over times (Hahn & Junge 1977). This is good for the farmer, but cause for concern for the atmospheric scientist.

Except for the production of N_2O by microorganisms, some atmospheric N_2O may be formed by lightning, as was suggested by Zipf (1975) and by Delwiche *et al.* (1977). Some N_2O is also produced during industrial ammonia oxidation, as was mentioned above, and during the combustion of fossil fuels (Pierotti & Rasmussen 1977).

The sinks of atmospheric N_2O are poorly known. The only major sink known so far is photolysis and chemical reaction with atomic oxygen, as already mentioned. For reasons which will be discussed, there have to be additional destruction processes somewhere in the troposphere, hydrosphere or pedosphere. Soil and sea water do not seem to come into question as major sinks of atmospheric N_2O because of problems with transport to anoxic sites where N_2O could be consumed by microorganisms. Assimilation of atmospheric N_2O by nitrogen fixing organisms is apparently also an inefficient sink, since N_2O has to compete with molecular nitrogen as a substrate, and molecular nitrogen is always present in large excess, even in water. Thus we are left with abiological destruction processes. Homogeneous reactions in the troposphere studied so far appear to destroy only small amounts of atmospheric N_2O. We know virtually nothing about heterogeneous destruction processes of N_2O in the troposphere or at the Earth's surface. Photolysis of N_2O absorbed on atmospheric particles is a conceivable destruction process, as was suggested by P. J. Ausloos (1977, private communication), but further proof is awaited.

Nitrous oxide balance of the troposphere

Available data show that in the troposphere N_2O is well mixed throughout the two hemispheres. Based on older measurements, the average tropospheric mixing ratio of N_2O was found to be 260 ± 25 parts/10^9 by volume (Hahn 1975; Hahn & Junge 1977). Recent measurements yielded values ranging from 280 to 340 parts $N_2O/10^9$ (see, for example, Rasmussen *et al.* 1976*a*, *b*; Yoshinari 1976; Weiss 1977; Singh 1977; Schmeltekopf *et al.* 1977; Brice & Penkett 1977; Cohen 1977). Although the precision of recent measurements is clearly better, the interlaboratory scatter of data has been still about 10 %. Apparently, this is primarily due to difficulties with calibration. Because of these difficulties and because the tropospheric N_2O level seems to be subject to long-term variations (Hahn & Junge 1977), the seemingly obvious conclusion that the tropospheric N_2O level has risen over the last decade cannot be drawn.

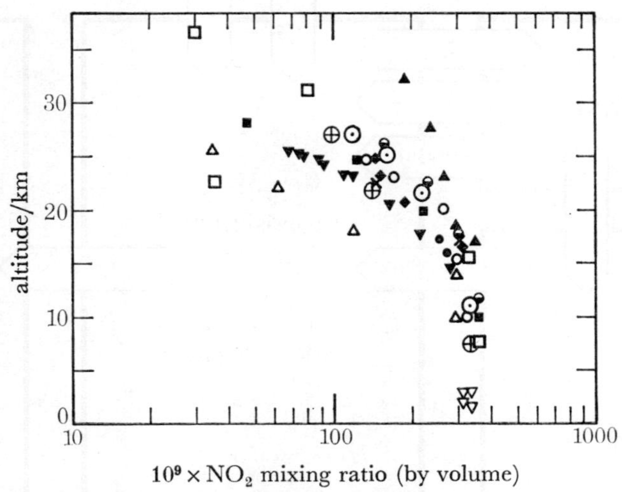

Figure 2. Atmospheric mixing ratio of N_2O at various locations and altitudes (figure taken from Schmeltekopf *et al.* 1977). ⊖, Wyoming, June 1975; △, Saskatchewan, Canada, September 1975; ▽, Colorado, September 1975; ●, Wyoming, December 1975; ▼, Antarctic, January 1976; ×, New Mexico, March 1976; ▲, Panama, April 1976; ⊕, Wyoming, May 1976; □, Alaska, May 1976; ■, Alaska, May 1976; ○, Wyoming, July 1976; ⊙, Wyoming, July 1976; ◆, Wyoming, August 1976.

In agreement with the older measurements, the more accurate modern data show virtually no vertical gradient of the N_2O mixing ratio in the troposphere and a pronounced decrease in the lower stratosphere. Vertical profiles recently obtained by Schmeltekopf and colleagues at five different latitudes are shown in figure 2 (taken from Schmeltekopf *et al.* 1977).

The profiles show that the N_2O mixing ratio is remarkably constant within the troposphere. In the stratosphere above 17 km, however, the variability is substantial, indicating a strong dependence on both horizontal transport within the stratosphere and photochemistry. Compared with the profiles obtained at high latitudes in both hemispheres, the profile measured in the tropics (Panama) indicates that vertical transport into the stratosphere is relatively large at low latitudes, a finding which is important for the calculation of mean global eddy diffusion coefficients used in one-dimensional transport models. Schmeltekopf and colleagues found values 1.5–3 times larger than Hunten (1975) who had no data from the tropics at that time.

With 260 parts/10^9 by volume, the tropospheric N_2O mass is 1700 Mt assuming an average height of the tropopause of 16 km in latitudes above 30° and of 11 km in latitudes below 30°.

Considering the precision of the measurements and the scatter data, the range of uncertainty is found to be 1700 ± 200 Mt. Based on recent data, a tropospheric N_2O mass of 2000 ± 200 Mt is obtained.

As was discussed in a recent paper (Hahn & Junge 1977), we think that there is sufficient evidence to assume that the turn-over time of tropospheric N_2O is relatively short. A value of 8 ± 4 years can be inferred from the older measurements. Rasmussen and coworkers' data (Rasmussen *et al.* 1976*a*, *b*), which represent the largest set of modern measurements within the troposphere, suggest a tropospheric turn-over of 10 ± 5 years.

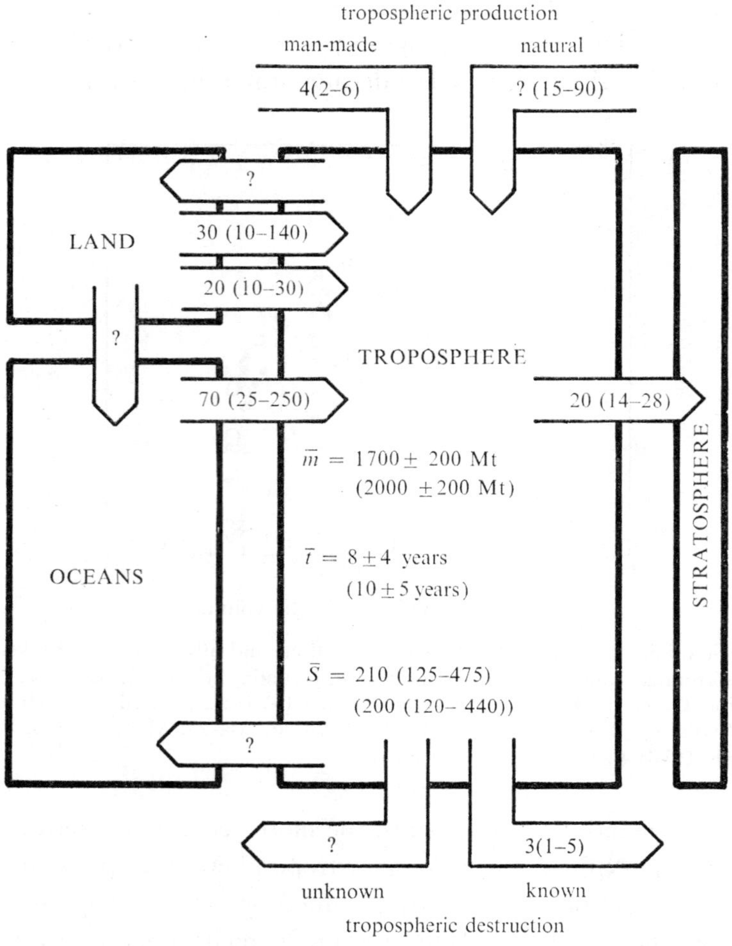

FIGURE 3. Nitrous oxide balance of the troposphere. Unless marked otherwise, all numbers given in the figure are in megatonnes N_2O per year. $\bar{m}$ denotes the tropospheric mass of N_2O averaged over a time period on the order of the tropospheric turn-over time $\bar{t}$. $\bar{s}$ is the total global source (sink) strength.

With a knowledge of the tropospheric mass ($\bar{m}$) and turn-over time ($\bar{t}$) of N_2O, the total flux ($\bar{s}$) into the troposphere and out of it, respectively, can be calculated ($\bar{m}/\bar{t} = \bar{s}$), if steady state is assumed. Considering, in addition, available data on N_2O in sea water and fresh water (Junge *et al.* 1971; Hahn 1974; Yoshinari 1976; Rasmussen *et al.* 1976; Hahn & Junge 1977), on N_2O in soil air (Arnold 1954; Albrecht *et al.* 1970; Schütz *et al.* 1970; Burford & Stefanson 1973; Rolston *et al.* 1976; Söderlund & Svensson 1976), and on N_2O in exhaust gases of industrial plants for the oxidation of ammonia (Leithe & Hofer 1968) and those of power plants (Pierotti & Rasmussen

1976), one can try to estimate individual global fluxes and construct an N_2O balance of the troposphere which appears to be the largest reservoir of N_2O. The result of such an effort is shown in figure 3.

Unless marked otherwise, all values given in the figure are in megatonnes N_2O per year. The mean total flux ($\bar{s}$) in or out of the troposphere considering a time span on the order of the turnover time ($\bar{t}$) is found to be 210 Mt/a† (with a range of uncertainty of 125–475 Mt/a) based on the older measurements and 200 Mt/a (with a range of uncertainty of 120–440 Mt/a) on the basis of recent measurements, two values which agree very well. The fluxes between the different reservoirs are indicated by large arrows.

The fluxes from the land surface into the troposphere have been somewhat arbitrarily divided in this graph into natural fluxes and fluxes due to the application of nitrogen fertilizers, although the processes which generate N_2O appear to be the same in both cases. The yield of N_2O, however may be different, (Hahn & Junge 1977). From measurements in soil air, a total global net N_2O production of natural soil of 25 Mt/a can be inferred. The range of uncertainty of 10–100 Mt/a reflects the difficulties of this assessment. The large uncertainty is due to the limited number of data on the one hand, and to the large variety of ecological conditions for the N_2O production in soil on the other hand. [In general, the microbial generation of N_2O depends on the temperature, pH and aeration (availability of free oxygen) of the habitat. In soil, the availability of free oxygen and the pH are determined not only by the soil type and structure, but also by the moisture content and the content of organic matter.] With respect to net N_2O production of fresh water, the situation is even worse. The only thing one can say at the moment is that this production rate must be somewhere between zero and 40 Mt/a (Hahn & Junge 1977). A first rough estimate would be about 5 Mt/a. This brings the natural production of the total land surface to about 30 Mt/a with a range of uncertainty of 10–140 Mt/a.

The net N_2O production of soil and fresh water with fertilizer nitrogen as a substrate may be estimated to be currently in the range of 10–30 Mt/a with 20 Mt/a as the most likely value. We arrive at this number when we consider the trend in fertilizer use over the last decade with about 40 Mt fertilizer nitrogen applied to soil in 1974. The range of 10–30 Mt/a corresponds to a yield factor of 0.2–0.6 for the nitrogen return to the atmosphere in the form of N_2O (Hahn & Junge 1977).

The most likely value for the net N_2O production of all oceans is estimated to be about 70 Mt/a (Hahn & June 1977). The estimate is based primarily on our measurements in the North Atlantic ocean using the stagnant film model (Bolin 1960; Kanwisher 1963) for the calculation of fluxes through the air–sea boundary layer. Assuming a mean thickness of the air–sea boundary layer of 60 μm (Broecker & Peng 1971, 1974), the value of 70 Mt/a would mean that the mean saturation of the surface water of the world ocean with N_2O should be between 150 and 160 % with respect to air (with 10 °C as the mean surface temperature of the world ocean). Values actually observed in the North Atlantic and in the Eastern Tropical Pacific range from 100 to 280 % saturation. In view of the fact that data on marine N_2O are still lacking from seemingly very productive areas such as the West African coast, the Indian ocean, and the Arabian Sea, and also from the South Atlantic, Central and South Pacific, and from the polar regions, the value of 70 Mt N_2O/a given in the graph should be taken with caution. The correct value should certainly however, be greater than 25 Mt/a.

† a = year.

Direct man-made sources of atmospheric N_2O seem to be relatively small at present. Pierotti & Rasmussen (1977) recently estimated the global release of N_2O to the atmosphere from the combustion of fossil fuels to be about 3.5 Mt/a. Considering additional input from other industrial processes, one arrives at about 4 Mt/a with 2–6 Mt/a as the range of uncertainty.

If the assessment by Dubin & Zipf (1977) is correct, lightning may be a significant source of tropospheric N_2O. From laboratory experiments, they inferred a figure of 15–90 Mt/a. Because of difficulties with the extrapolation of measurements carried out in the laboratory to the real atmosphere, it is felt that more experiments and measurements in the field are necessary to validate these numbers.

As already mentioned, the sinks of tropospheric N_2O are not too well known. The only sink one can be fairly sure of is in the stratosphere. Assessments by Junge (1972), Crutzen (1976), McElroy *et al.* (1977) and Schmeltekopf *et al.* (1977) cover the range from 14 to 28 Mt N_2O/a. In figure 3, 20 Mt/a was considered to be the most likely value. Except for a minute amount of about 3 Mt/a which may be destroyed by homogeneous reactions in the troposphere, no other sinks of atmospheric N_2O are known. This, however, should not lead us to conclude that there are no other sinks and that the tropospheric turn-over time of N_2O must therefore be on the order of 100 years. The variation of the tropospheric N_2O level with time strongly suggests a relatively short turn-over time as given in figure 3. At the present state of knowledge, it is hard to understand how the oceans and the soil can act as significant sinks of atmospheric N_2O. Assuming steady state in the troposphere, additional sinks totalling about 180 Mt N_2O/a remain to be found. As mentioned above, one should take a closer look into the possibility of N_2O destruction by heterogeneous reactions in the troposphere itself or at the soil surface.

References (Hahn)

Albrecht, B., Junge, C. & Zakosek, H. 1970 *Z. PflErnähr. Bodenk.* **125**, 205–211.
Arnold, P. W. 1954 *J. Soil Sci.* **5**, 116–128.
Bolin, B. 1960 *Tellus* **12**, 274–281.
Brice, J. A. & Penkett, S. A. 1977 *Measurements of nitrous oxide in the atmosphere via E.C.D. gas chromatography at Harwell.* U.K. Atomic Energy Authority, ERE–R8569. London: H.M.S.O.
Broecker, W. S. & Peng, T.-H. 1971 *Earth planet. Sci. Lett.* **11**, 99–108.
Broecker, W. S. & Peng, T.-H. 1974 *Tellus* **26**, 21–35.
Burford, J. R. & Stefanson, R. C. 1973 *Soil Biol. Biochem.* **5**, 133–141.
Cohen, Y. 1977 *Anal. Chem.* **49**, 1235–1237.
C.I.S.C. 1976 *Halocarbons: environmental effects of chlorofluoromethane release.* Washington, D. C.: Committee on Impacts of Stratospheric Change, National Academy of Sciences.
Crutzen, P. J. 1970 *Q. Jl R. met. Soc.* **96**, 320–325.
Crutzen, P. J. 1976 *Geophys. Res. Lett.* **3**, 169–172.
Delwiche, C. C., Bissell, S. & Virginia, R. 1977 In *Proc. Kearney Foundation Workshop on 'Nitrogen in relation to environmental quality'*, Lake Arrowhead, California, U.S.A., 31 January to 3 February. Academic Press.
Dubin, M. & Zipf, E. P. 1977 *Trans. Am. geophys. Un.* **58**, 646 (SA 131).
Ehhalt, D. H., Heidt, L. E., Lueb, R. H. & Roper, N. 1974 In *Proc. Third Conf. Climatic Impact Assess. Progr.* pp. 153–160. Washington, D.C.: U.S. Department of Transport.
Goody, R. M. 1969 *Planet. Space Sci.* **17**, 1319–1320.
Hahn, J. 1974 *Tellus* **26**, 160–168.
Hahn, J. 1975 *Meteor. Forsch.-Ergebn* A **16**, 1–14.
Hahn, J. & Junge, C. 1977 *Z. Naturf.* A **32**, 190–214.
Harries, J. E., Birch, J. R., Fleming, J. W., Stone, N. W. B., Moss, D. G., Swann, N. R. W. & Neill, G. F. 1974 *Proc. Third Conf. Climatic Impact Assess. Progr.*, pp. 197–212. Washington, D.C.: U.S. Department of Transport.
Hunten, D. M. 1975 *Proc. nat. Acad. Sci. U.S.A.* **72**, 4711–4715.
Johnston, H. S. 1971 *Science, N.Y.* **173**, 517–519.
Junge, C., Bockholt, B., Schütz, K. & Beck, R. 1971 *Meteor Forsch.-Ergebn.* B **6**, 1–11.
Junge, C. 1972 *Q. Jl R. met. Soc.* **98**, 711–729.

Kanwisher, J. 1963 *Deep Sea Res.* **10**, 195–207.
LaHue, M. D., Axelrod, H. D. & Lodge, J. P. 1970 *J. geophys. Res.* **75**, 2922–2926.
Leithe, W. & Hofer, A. 1968 *Allg. prakt. Chem.* **19**, 78–79.
McElroy, M. B., Wofsy, S. C. & Yung, Y. L. 1977 *Phil. Trans. R. Soc. Lond.* **277**, 159–181.
Murcray, D. G., Goldman, A., Murcray, F. H., Williams, W. J., Brooks, J. N. & Barker, D. B. 1973 *Proc. Sec. Conf. Climatic Impact Assess. Progr.*, pp. 86–98. Washington, D.C.: U.S. Department of Transport.
Pierotti, D. & Rasmussen, R. A. 1976 *Geophys. Res. Lett.* **3**, 265–267.
Rasmussen, R. A., Pierotti, D., Krasnec, J. & Halter, B. 1976*a* *Report on the cruise of the Alpha-Helix research vessel, March* 5–20, 1976. Air resources section, College of Engineering, Washington State University, Pullman, Washington, U.S.A.
Rasmussen, R. A., Krasnec, J. & Pierotti, D. 1976*b* *Geophys. Res. Lett.* **3**, 615–618.
Rolston, D. E., Fried, M. & Goldhamer, D. A. 1976 *Soil Sci. Soc. Am. J.* **40**, 259–266.
Schmeltekopf, A. L., Albritton, D. L., Crutzen, P. J., Goldan, P. D., Harrop, W. J., Henderson, W. R., McAfee, J. R., McFarland, M., Schiff, H. I., Thompson, T. L., Hofmann, D. J. & Kjome, N. T. 1977 *J. atmos. Sci.* **34**, 729–736.
Schütz, K., Junge, C., Beck, R. & Albrecht, B. 1970 *J. geophys. Res.* **75**, 2230–2246.
Singh, H. B. 1977 *Geophys. Res. Lett.* **4**, 101–104.
Söderlund, R. & Svensson, B. H. (eds) 1976 *Scope report* no. 7 and *Ecological Bulletins* no. 22. Stockholm: Swedish Natural Science Research Council.
Weiss, R. F. 1977 Unpublished work. Scripps Oceanographic Inst., La Jolla, California, USA.
Yoshinari, T. 1976 *Mar. Chem.* **4**, 189–202.
Zipf, E. P. 1975 Laboratory studies on the synthesis of N_2O and NO_2, and on the destruction of CF_2Cl_2 by lightning. Paper presented at the A.G.U. Fall Annual meeting, San Francisco, California, U.S.A.

Discussion

B. A. THRUSH, F.R.S. (*University of Cambridge, Department of Physical Chemistry, Lensfield Road, Cambridge CB2 1EP, U.K.*). It has been suggested that the increased acidity of rain might increase the ratio of nitrous oxide to nitrogen in denitrification processes and that this could provide a positive feedback mechanism enhancing the ozone depletion which is expected to arise from greatly increased production of nitrate fertilizers. Could Dr Hahn comment on this suggestion?

J. HAHN. Although it appears to be fairly well established that with decreasing pH the ratio $N_2O:N_2$ in the products of bacterial denitrification increases, the effect of a decrease in the pH of rain water on the microbiological production of atmospheric N_2O is hard to assess. There is the buffer capacity of soil and water bodies on one hand; on the other there is the observation that the rates of both nitrification and denitrification decrease with decreasing pH in the microbial habitat.

Soils with limited buffer capacity are found in areas with granitic bedrock and are common in Canada and in Scandinavia. On a global scale, however, soils derived from sedimentary bedrock containing large amounts of clay minerals and changing amounts of carbonates and humus predominate which have a very good buffer capacity. Soils typical of the humid parts of the temperate zones are the podzols and their relatives. Abundant vegetation makes the soil water acidic anyway (pH reaching 4.0–4.5 in the clay portion of the soil and as low as 3.5 in the humus) so that acidic rain water, as long as it is not pure sulphuric or hydrochloric acid, would not change too much in these zones as far as the acidity of the microbial habitats is concerned.

Even if the rain does change the pH in the microbial habitats, the effect would be that the nitrogen cycle in the systems affected by a decreasing pH would slow down more and more, very probably with decreasing rather than increasing net N_2O production rates. Fixed nitrogen which is added to the soil in such areas and which is not incorporated into plants would either be trapped in the soil organic matter or leached out and carried off to where conditions are more favourable for nitrification and denitrification (ultimately to the sea with its extremely large buffering potential).

T. Nash (*Microbiological Research Establishment, Porton Down, Wilts., U.K.*). I suggest that the 'unknown main sink' for nitrous oxide, mentioned by Dr Hahn, is the photosynthetic cycle for the reduction of carbon dioxide. Nitrous oxide should be able to enter the cycle by being isoelectronic with carbon dioxide. There is a precedent here in the form of the acetylene reaction in nitrogen fixation. I further suggest that carbonic anhydrase should be able to hydrate nitrous oxide reversibly to hyponitrite, permitting biological reduction to ammonia.

J. Hahn. The possibility that photosynthesis and/or nitrogen fixation could act as main sinks of atmospheric N_2O has been considered. However, there is as yet no support for this idea.

In higher plants, no evidence for any active assimilation of N_2O could be detected. As far as microorganisms are concerned, studies have shown that *Azotobacter vinelandii* is able to metabolize N_2O slowly and that the N_2O nitrogen subsequently appears in the microbial protoplasm. Similar observations have been made in laboratory studies with certain soil–plant combinations. If one accepts the premise that these observations indicate a real assimilatory process and not a combination of local denitrification and subsequent N_2 fixation, one must conclude that N_2O – if present in sufficiently large mixing ratio, i.e. in the percentage range rather than in parts/10^9 – is slowly taken up and reduced along the reaction pathway used in photosynthesis and/or along the reaction pathway used in nitrogen fixation. Nitrogen fixation would be more plausible, since it has been that N_2O acts as a competitive inhibitor of N_2 uptake by microorganisms such as *Azotobacter* and *Clostridium*. Regardless of which is correct, a sink of atmospheric N_2O on the basis of photosynthesis and nitrogen fixation can only be minor judging from what we know at present.

Phil. Trans. R. Soc. Lond. A. **290**, 505–514 (1979)
Printed in Great Britain

Aspects of the chemistry of ozone depletion

By B. A. Thrush, F.R.S.

University of Cambridge, Department of Physical Chemistry, Lensfield Road, Cambridge CB2 1EP, U.K.

The conversion of stratospheric ozone and atomic oxygen back to molecular oxygen is catalysed by a number of atmospheric trace species such as nitrogen oxides (NO, NO_2), halogen (Cl, ClO) and hydrogen (H, HO, HO_2) radicals. The concentrations of these species are interrelated by a complex series of chemical reactions where those of hydroperoxy (HO_2) are the least well understood. Recent direct laboratory studies of the reactions of HO_2 have a considerable effect on estimates of ozone depletion by anthropogenic nitrogen oxides and halogen compounds.

Introduction

An aspect of atmospheric pollution which has caught the public attention in recent years is depletion of stratospheric ozone. One effect of this would be an increase in the amounts of ultraviolet light between 290 and 320 nm (u.v.-B) reaching ground level. Such light is biologically active and prolonged exposure to it can cause skin cancer in susceptible individuals (Booker 1975). The considerable annual and longer-term fluctuations in the ozone cover make it difficult to observe the relatively slow long-term changes which stratospheric pollutants are expected to produce. For this reason, assessment of the probable impact of stratospheric pollutants is based on modelling calculations. For such calculations we need to know the tropospheric sources and sinks of the species involved; nitrous oxide, which was discussed in the previous paper (Hahn 1979, this volume) is an important example. The problems of modelling the processes which transport the pollutants and their degradation products into and out of the stratosphere have already been discussed by Tuck (1979, this volume). Here, we consider the chemical processes which destroy stratospheric ozone, particularly in relation to the uncertainties of predicting ozone depletions and atmospheric measurements of trace species.

Ozone chemistry

Ozone is thinly distributed over the stratosphere, which is the region of temperature inversion between the tropopause at 10–20 km and the stratopause at 50 km. Its mixing ratio (mole fraction) barely exceeds 10^{-6} and the total amount of ozone in the Earth's atmosphere corresponds to a column height of about 3 mm at s.t.p. (figure 1).

The basic photochemistry of an oxygen atmosphere was established by Chapman in 1930:

$$O_2 + h\nu = O + O, \qquad (1)$$

$$O + O_2 + M = O_3 + M, \qquad (2)$$

$$O_3 + h\nu = O_2 + O, \qquad (3)$$

$$O + O_3 = O_2 + O_2. \qquad (4)$$

Oxygen is dissociated by light with wavelengths below 242 nm to produce oxygen atoms which combine with molecular oxygen to form ozone. In the stratosphere ozone absorbs sunlight much more strongly than does oxygen. As a result the interconversion of the 'odd oxygen' species O and O_3 by (2) and (3) is orders of magnitude faster than their formation by (1) or removal by (4). Laboratory measurements (Davis *et al.* 1973) of the Arrhenius parameters of (4), $k_4 = 1.9 \times 10^{-11} \exp(-2300/T)\ \mathrm{cm^3\,s^{-1}}$, show it to have an unusually low pre-exponential factor and high activation energy for such an exothermic process, making it slow at stratospheric temperatures.

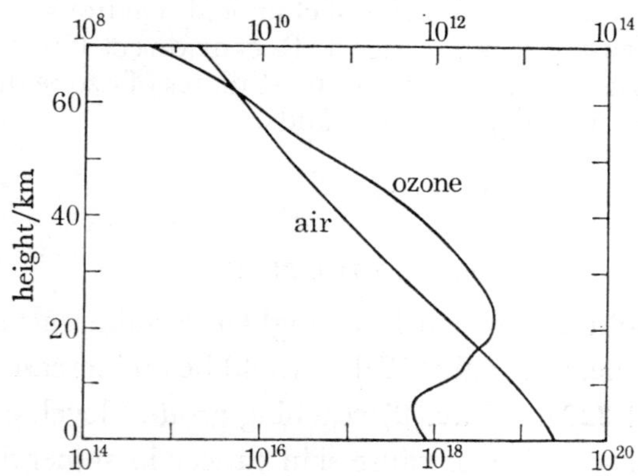

FIGURE 1. Profiles of ozone density and air density in mid-latitudes. Horizontal scales are displaced by a factor of 10^6 to give ozone in parts/10^6 from their separation (Booker 1975).

An assessment of the total rate of formation of odd oxygen in the atmosphere and its removal by (4) and by slow transport to the surface based on the observed ozone distribution shows that these processes can only account for about 20 % of the actual rate of removal of odd oxygen (ozone) (Johnston 1974). Other chemical processes must therefore account for the bulk of ozone destruction and their basic mechanisms can be simplistically represented by a chain mechanism which achieves the same result as reaction (4):

$$X + O_3 = XO + O_2 \qquad (5)$$

$$XO + O = X + O_2 \qquad (6)$$

$$\overline{\qquad\qquad\qquad\qquad}$$

$$O + O_3 = O_2 + O_2.$$

The condition that both (5) and (6) should be exothermic is

$$100\ \mathrm{kJ\,mol^{-1}} < D(X-O) < 485\ \mathrm{kJ\,mol^{-1}}$$

which is clearly satisfied by many gaseous species X. The most important of these, particularly in the lower stratosphere, involves X = NO (Crutzen 1970; Johnston 1971), while in the upper stratosphere and mesosphere, cycles with X = HO and X = H contribute significantly.

Halogen atoms can also play the rôle of X, as has been shown by recent concern over the fate of chlorine liberated in the decomposition of aerosol propellants and halogenated solvents released into the atmosphere.

However, an estimate of the rate of ozone removal depends not merely on the laboratory determination of the rates of (5) and (6) for various species X. It also depends on the rates of the processes forming and removing X or XO in the stratosphere (chemistry and transport) and on processes which interconvert X and XO without ozone removal. These are well illustrated for the NO_x and ClX cycles by figure 2. For the former we can distinguish three types of cycle:

ozone destruction

$$NO + O_3 = NO_2 + O_2 \quad (7)$$
$$O + NO_2 = NO + O_2 \quad (8)$$
$$\overline{O + O_3 = O_2 + O_2}$$

'no effect' cycle

$$NO + O_3 = NO_2 + O_2 \quad (7)$$
$$NO_2 + h\nu = NO + O \quad (9)$$
$$\overline{O_3 = O_2 + O}$$

holding cycle

$$HO + NO_2 + M = HONO_2 + M \quad (10)$$
$$HNO_3 + h\nu = HO + NO_2 \quad (11)$$
$$HO + HNO_3 = H_2O + NO_3 (\xrightarrow{h\nu} NO_2) \quad (12)$$

Here (8) controls the rate of odd oxygen destruction because it competes with the 'no effect' process (9) and with (10) which results in the formation of the relatively long-lived nitric acid molecule.

For ClX the 'no effect' cycle involves nitrogen oxides:

ozone destruction

$$Cl + O_3 = ClO + O_2 \quad (13)$$
$$O + ClO = Cl + O_2 \quad (14)$$
$$\overline{O + O_3 = O_2 + O_2}$$

'no effect' cycle

$$ClO + NO = Cl + NO_2 \quad (15)$$
$$NO_2 + h\nu = NO + O \quad (9)$$
$$Cl + O_3 = ClO + O_2 \quad (13)$$
$$\overline{O_3 = O_2 + O}$$

and two holding processes have been identified, the main one involving HO,

$$Cl + CH_4 = CH_3 + HCl, \quad (16)$$
$$HO + HCl = H_2O + Cl, \quad (17)$$

the other involving nitrogen oxides,

$$ClO + NO_2 + M = ClONO_2 + M, \quad (18)$$

although chlorine nitrate has not yet been detected in the stratosphere.

Thus the rates of destruction of odd oxygen by the rate-determining processes (8) and (14) are governed by the proportions of NO_x (NO, NO_2, HNO_3) and ClX (Cl, ClO, HCl) which are present as NO_2 and ClO respectively. In particular, HO plays a crucial rôle in both systems, where it forms the holding species HNO_3 but removes the holding species HCl. The reactions which form, interconvert and remove the HO_x species H, HO and HO_2 are illustrated by a Nicolet diagram in figure 3. Although the rate of production of HO_x species by the reactions with water, methane and hydrogen of the excited oxygen atoms, $O(^1D)$, produced by ozone photolysis is comparatively well known, the rates of interconversion of H, HO and HO_2 and their removal, principally by the reaction

$$HO + HO_2 = H_2O + O_2, \quad (19)$$

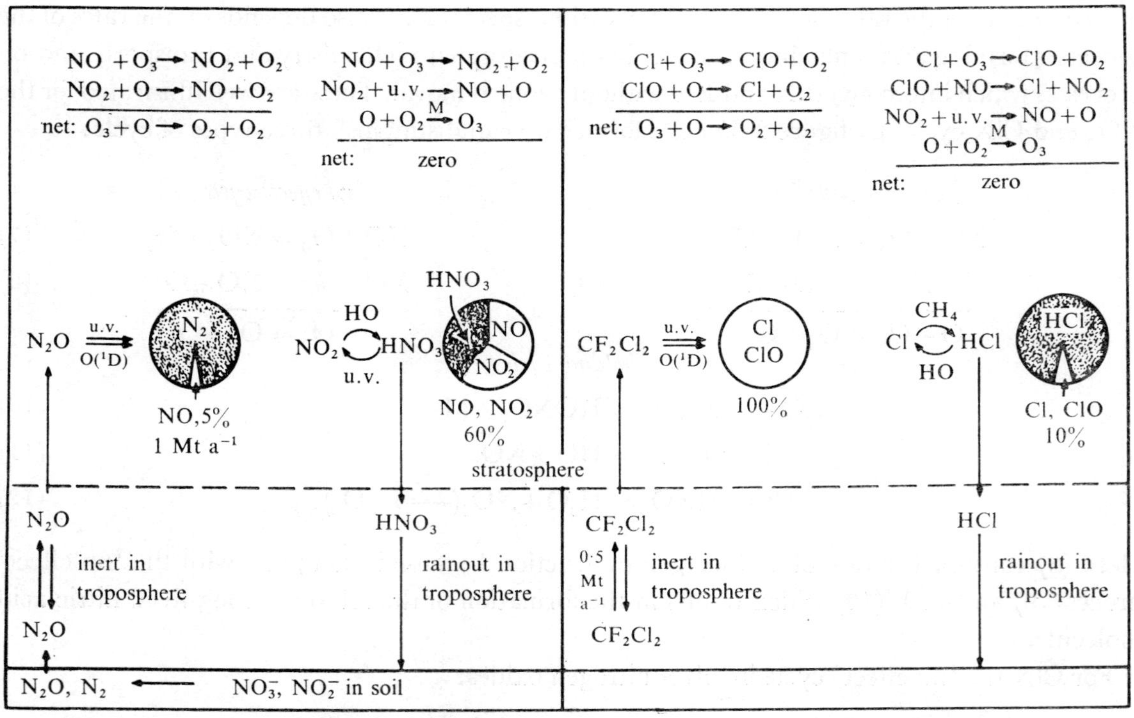

FIGURE 2. Comparison of the nitrogen oxide and chlorine cycles for ozone removal, illustrating the relative proportions of active and inactive species (Johnston 1975).

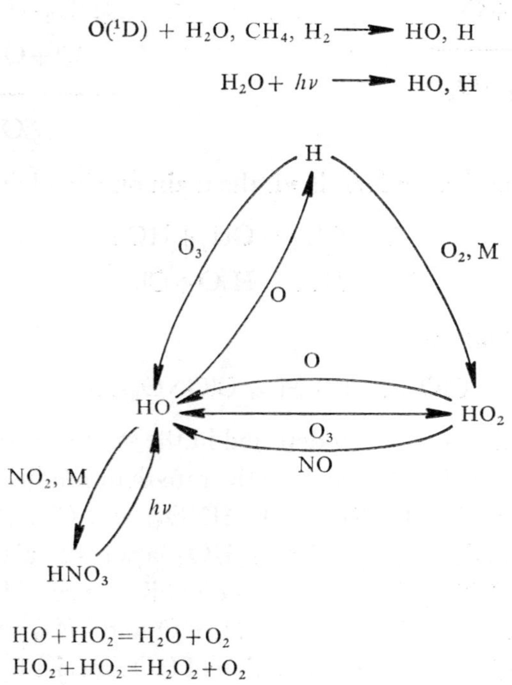

FIGURE 3. Sources, dominant interconversion processes and removal processes of hydrogen radicals (H, HO, HO_2) in the stratosphere.

contain major uncertainties, largely associated with the difficulty of making direct laboratory studies of the reactions of HO_2. Some recent investigations of these reactions and their consequences in atmospheric chemistry are considered below.

Reactions of hydroperoxy (HO_2)

Most laboratory studies of reactions of stratospheric importance use either discharge flow systems or flash photolysis with highly sensitive specific methods for determining the concentrations of active species such as chemiluminescence (O, NO), resonance fluorescence (H, O, Cl, HO) absorption spectroscopy (ClO, O_3) or mass spectrometry (O, Cl, ClO, etc.). Unfortunately, HO_2 cannot be detected by chemiluminescence. Its electronic spectrum consists of weak bands near 1.2 μm and featureless continuous absorption around 200 nm where many other species absorb.

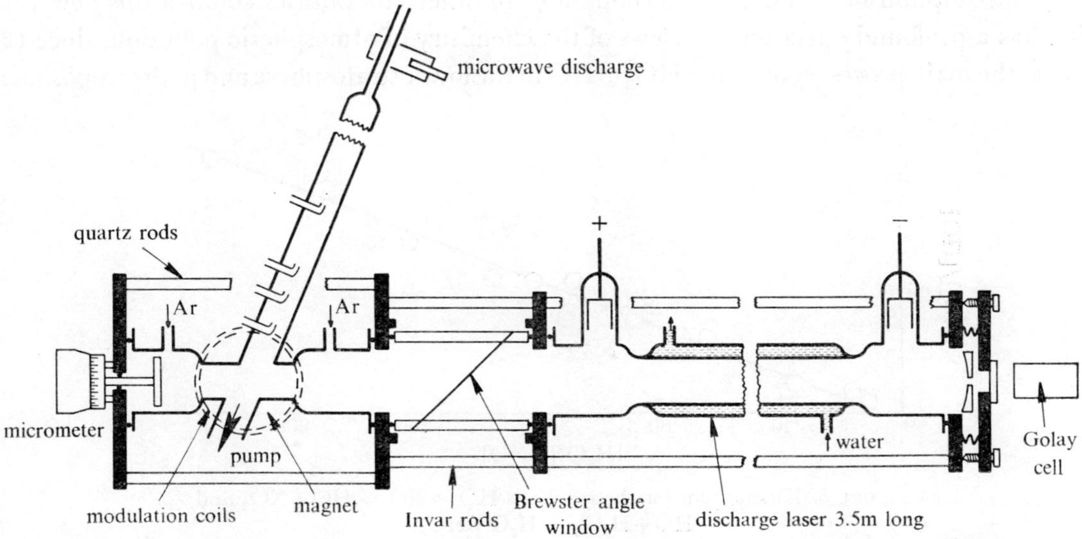

FIGURE 4. Laser magnetic resonance spectrometer for free radical studies (Burrows *et al.* 1977).

These spectra are unsuitable for resonance fluorescence and cause difficulties in attribution of absorption measurements. With mass spectrometry, calibration can cause difficulties particularly as the ion HO_2^+ is readily formed from other species. Recently, laser magnetic resonance spectroscopy in which a strong magnetic field is used to bring Zeeman components of the rotational transitions of a free radical into resonance with a far infrared laser has provided a highly specific method for the detection and study of such radicals as HO_2. Furthermore, the problem of calibration is minimized because the intensity of the spectrum depends directly on the square of the dipole moment of the species rather than on the dipole moment derivative (vibration spectrum) or an electronic transition moment.

Figure 4 shows the basic arrangement of a discharge-flow system for studying atomic and free radical reactions coupled to a laser magnetic resonance spectrometer based on the 118.6 μm water vapour line with which it is possible to detect HO_2, HO and NO_2. With this apparatus we have been able to make the first measurement of the rate of the reaction.

$$O + HO_2 = HO + O_2, \qquad (20)$$

and the first direct measurement of the rate of coefficient of

$$HO + HO_2 = H_2O + O_2. \tag{19}$$

Both these rate coefficients (Burrows *et al.* 1977) were fairly close to values used by atmospheric modellers but a more recent determination of the rate of the reaction

$$HO_2 + NO = HO + NO_2, \tag{21}$$

in which nitric oxide was added to HO_2 radicals produced by a Tesla discharge through hydrogen peroxide (figure 5), yielded

$$k_{21} = (8.2 \pm 1.2) \times 10^{-12}\,cm^3\,s^{-1},$$

which is over twenty times the value normally used in stratospheric modelling (Burrows *et al.* 1978). Howard & Evenson (1977) have obtained a similar value in a variant of the same technique, and unpublished indirect determinations in other laboratories confirm this new value which has a profound effect on our views of the chemistry of atmospheric pollution, since (21) becomes the main process converting HO_2 to HO in the lower stratosphere and in the troposphere.

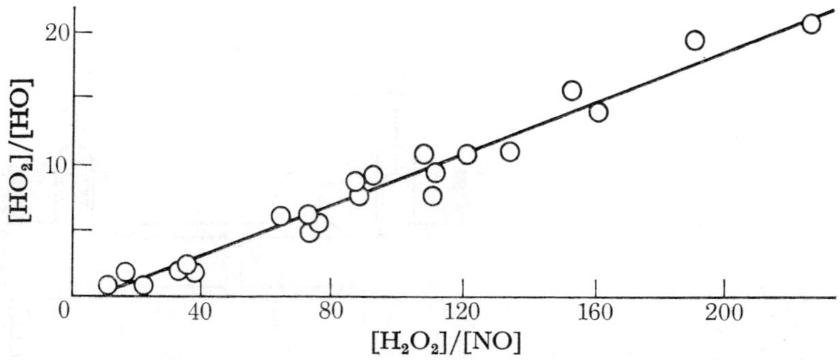

FIGURE 5. Kinetic plot for the reactions $HO_2 + NO = HO + NO_2$ and $HO + H_2O_2 = H_2O + HO_2$.

In particular, interconversion of HO_2 and HO in the lower stratosphere is now believed to be controlled by the 'no effect' cycle,

$$HO_2 + NO = HO + NO_2, \tag{21}$$

$$HO + O_3 = HO_2 + O_2, \tag{22}$$

$$NO_2 + h\nu = NO + O, \tag{9}$$

rather than by the ozone destruction cycle

$$HO_2 + O_3 = HO + O_2 + O_2, \tag{23}$$

$$HO + O_3 = HO_2 + O_2. \tag{22}$$

It also increases the calculated ratio of HO to HO_2 in the upper stratosphere where the corresponding ozone destruction cycle is

$$O + HO_2 = HO + O_2, \tag{20}$$

$$HO + O_3 = HO_2 + O_2. \tag{22}$$

Ozone depletion

These improvements in our knowledge of reaction rates for HO_2 have considerable effects on estimates of the depletion of stratospheric ozone by various pollutants.

For chlorofluoromethanes and other chlorocarbons, where the maximum ozone depletion is predicted to occur around 30 km altitude, the calculated depletions (National Academy of Sciences (Gutowsky 1976)) are increased by about one half. This arises because increased HO concentrations convert more HCl into active Cl and ClO and because the removal of NO by (21) reduces the rate of (15) which initiates a 'no effect' cycle.

Ozone removal by nitrogen oxides occurs predominantly in the lower stratosphere, and the predicted effect of NO and NO_2 depends on how close to the tropopause they are released. The nitrogen oxides from Concorde and high-altitude subsonic aircraft such as the Boeing 747SP which operate at altitudes up to 18 km are now predicted to enhance ozone slightly. This arises because:

(i) the HO, HO_2 depletion cycle of (22) and (23) is partly replaced by the 'no effect cycle' of (22) (21) and (9);

(ii) a higher concentration of HO makes reaction (12) between nitric acid and HO compete significantly with the photolysis of nitric acid in the lower stratosphere. Increases in nitrogen oxide levels then causes HO_x removal by reaction (19),

$$HO+HO_2 = H_2O+O_2, \tag{19}$$

to be replaced by the following cycle which generates odd oxygen:

$$HO_2+NO = HO+NO_2 \tag{21}$$

$$HO+NO_2+M = HNO_3+M \tag{10}$$

$$HO+HNO_3 = H_2O+NO_3 \tag{12}$$

$$NO_3+h\nu = NO_2+O \tag{24}$$

$$NO_2+h\nu = NO+O \tag{9}$$

$$\overline{HO+HO_2 = H_2O+O+O.}$$

With a high value of k_{21}, the oxidation of CO to CO_2 by HO_x species becomes an odd oxygen generating cycle,

$$HO+CO = CO_2+H \tag{25}$$

$$H+O_2+M = HO_2+M \tag{26}$$

$$HO_2+NO = HO+NO_2 \tag{21}$$

$$NO_2+h\nu = NO+O \tag{9}$$

$$\overline{CO+O_2 = CO_2+O,}$$

instead of an odd oxygen removing cycle,

$$HO+CO = CO_2+H \tag{25}$$

$$H+O_2+M = HO_2+M \tag{26}$$

$$O+HO_2 = HO+O_2 \tag{20}$$

$$[\text{or } HO_2+O_3 = HO+O_2+O_2] \tag{23}$$

$$\overline{CO+O = CO_2.}$$

Release of NO and NO_2 above 20 km which might occur from future supersonic aircraft and occurs from the reaction

$$O(^1D) + N_2O = NO + NO, \tag{27}$$

which is the dominant natural source of stratosphere NO and NO_2, is still predicted to deplete ozone somewhat, as HNO_3 is formed more slowly and photolysed more rapidly at these altitudes. Thus these new results reduce rather than eliminate predictions of ozone depletion by nitrous oxide derived from bacterial denitrification of nitrate fertilizers, the use of which is increasing rapidly. However, as Hahn (1979, this volume) has pointed out, we do not yet know enough about the natural N_2O cycle to predict the increase in atmospheric N_2O due to this source which would be needed to predict any ozone depletion.

Modelling the chemistry of the stratosphere involves about 100 chemical reactions. Although a better knowledge of the rates of many of these processes is unlikely to affect significantly our views on ozone chemistry, future laboratory studies, particularly of the reactions of HO_2, may well yield other surprises. For instance, the rates of (23) and (28),

$$HO_2 + HO_2 = H_2O_2 + O_2, \tag{28}$$

are not well established, and the recent observation of the formation of pernitric acid from the combination of HO_2 and NO_2 (Niki *et al.* 1977) and its attempted detection in the stratosphere by D.G. Murcray (unpublished results) point out directions for future investigations. With the simultaneous measurements of a number of trace species in the stratosphere now being planned in a number of laboratories we should soon have quantitative tests of the accuracy of our predictions of the rates of the various chemical cycles involving stratospheric ozone.

References (Thrush)

Booker, H. G. 1975 *Environmental impact of stratospheric flight.* Washington, D.C.: Climatic Impact Committee, National Academy of Sciences.

Burrows, J. P., Harris, G. W. & Thrush, B. A. 1977 *Nature, Lond.* **267**, 233.

Burrows, J. P., Cliff, D. I., Harris, G. W., Thrush, B. A. & Wilkinson, J. P. T. 1978 To be published.

Chapman, S. 1930 *Mem. R. met. Soc.* **3**, 103.

Crutzen, P. 1970 *Q. Jl R. met. Soc.* **96**, 320.

Davis, D. D., Wong, W. & Lephardt, J. 1973 *Chem. Phys. Lett.* **22**, 273.

Gutowsky, H. S. 1976 *Halocarbons: effects on stratospheric ozone.* Washington, D.C.: Panel on Atmospheric Chemistry of the Committee on Impacts of Stratospheric Change, National Academy of Sciences.

Hahn, J. 1979 *Phil. Trans. R. Soc. Lond.* A **290**, 495–504 (this volume).

Howard, C. J. & Evenson, K. M. 1978 *Geophys. Res. Lett.* **4**, 437.

Johnston, H. S. 1971 *Science, N.Y.* **173**, 517.

Johnston, H. S. 1974 *Acta astron.* **1**, 135.

Johnston, H. S. 1975 *A. Rev. phys. Chem.* **26**, 315.

Niki, H., Maker, P. D., Savage, C. M. & Breitenbach, L. P 1977 *Chem. Phys. Lett.* **45**, 564.

Tuck, A. F. 1979 *Phil. Trans. R. Soc. Lond.* A **290**, 477–494 (this volume).

Discussion

A. F. Tuck (*Meteorological Office, Bracknell, Berkshire, U.K.*). What is your opinion of the possible importance of pernitric acid, HO_2NO_2, and hypochlorous acid, HOCl, in stratospheric chemical kinetics?

B. A. Thrush. In the gas phase, Cl and HO normally exhibit very similar reactivities and laboratory studies show that this similarity extends to ClO and HOO, notably in their rapid reactions with NO and in their three-body combination with NO_2 to form respectively chlorine nitrate ($ClONO_2$) and pernitric acid ($HOONO_2$). The atmospheric photolysis of both these molecules is thought to be slow and their reactivity towards free radicals relatively low. On this basis, one would expect pernitric acid to be more important than chlorine nitrate at the present time.

We cannot assess the rôle of hypochlorous acid (HOCl) until we have more data on its absorption spectrum and reactions. It is closely related chemically and probably spectroscopically to hydrogen peroxide, which has not yet been detected in the stratosphere. As hydrogen peroxide is expected to be more abundant than hypochlorous acid, the relation between its predicted and observed concentrations would be useful in assessing the importance of hypochlorous acid.

B. J. Mason, Treas.R.S. (*Meteorological Office, Bracknell, Berkshire, U.K.*). The running of a complex three-dimensional dynamical model of the stratosphere, such as that developed in the Meteorological office and incorporating the full ozone photochemistry, would be very expensive in computer time. In view of the present uncertainties concerning the concentrations of key species and the rate constants of many of the reactions, I should like to ask Dr Thrush whether, in his opinion, such computations are justified at present. My own inclination would be to support Dr Tuck in incorporating the essential transport properties culled from the three-dimensional meteorological model into a one-dimensional model of the 'full' photochemistry and use this to test the sensitivity of ozone production and destruction to changes in the concentrations of the important species, to variations in reaction rates, and to the inclusion of new reactions.

B. A. Thrush. The total cost of the laboratory determination of the rate coefficient for a reaction of HO_2 is about £10 000 and until more rate coefficients are accurately known, I believe it is more appropriate to work on improving techniques for chemical modelling than to undertake very large computations on stratospheric chemistry at present. An important intermediate stage is the simultaneous measurement of concentrations of related trace species under conditions where their relative concentrations are determined by chemical processes and not by transport. Such measurements will test the adequacy and accuracy of the chemical model employed.

T. M. Sugden, F.R.S. What is the state of our knowledge of the total flux of chlorine to the stratosphere?

B. A. Thrush. The measured vertical profiles for F-11 and F-12 agree well with the calculated ones indicating that their fluxes into the stratosphere can be predicted with reasonable accuracy. However, the dominant chlorine species in the troposphere is methyl chloride for which Lovelock has suggested that tropical slash-and-burn agriculture is a major source. As much of the transport from the troposphere into the stratosphere is believed to occur via tropical storms which penetrate the tropopause, the flux in methyl chloride into the stratosphere must be highly uncertain.

Although Anderson's recent measurements of ClO at altitudes between 25 and 40 km give vertical profiles which agree well with modelling calculations, they correspond to total chlorine concentrations of 5 parts/10^9 by volume or more, whereas the more numerous measurements of HCl mainly at lower altitudes correspond to values around 2 parts/10^9. There is no obvious flaw in Anderson's measurements, but they are hard to reconcile with the observed ozone concentrations and with the measured amounts of chlorine compounds at lower altitudes. However, they may point to a lack of understanding of the stratospheric chlorine cycle.

G. Diprose (*I.C.I. Limited, Mond Division, Winnington, Northwich CW8 4DJ, U.K.*). Our knowledge and understanding of fluxes and concentration profiles of chlorinated species in the stratosphere is still very far from complete and there are a number of anomalies such as the decrease in HCl concentration measured above 35 km and the very high measurements of ClO reported by Anderson still to be explained. In addition we have the possibility that HOCl may be a stratospheric species of importance and that there may be other species still awaiting recognition.

All this puts at a premium a reliable method for the determination of total chlorine in the stratosphere. Comparison of such measurements with simultaneous determinations of individual chlorinated species can act as a check on the measurements themselves and can give some indication if there is a hitherto unrecognized species of importance in the stratosphere and also help to put the relative fluxes of natural and man-made chlorine-containing compounds into their correct perspective.

Concluding remarks

B. A. Thrush. My paper stressed the importance to stratospheric pollution of knowing the concentrations of hydroxyl radicals in the stratosphere. Greater concentrations would decrease ozone removal by nitrogen oxides but increase ozone removal by halogen compounds. Dr Swift-Hook and others have pointed out the need for measurement of free radical concentrations in the troposphere, and in the case of hydroxyl this is particularly important, since this species plays a large rôle in photochemical smog, in oxidizing sulphur dioxide, in converting carbon monoxide to carbon dioxide, and in removing methyl chloride and other partly halogenated hydrocarbons. It is interesting to note that the concentration of this important scavenger is much higher in urban than in rural atmospheres.

Phil. Trans. R. Soc. Lond. A. **290**, 515–521 (1979)
Printed in Great Britain

Greenhouse effects of some atmospheric constituents

By J. T. Houghton, F.R.S.
Department of Atmospheric Physics, Clarendon Laboratory, Oxford, U.K.

The present state of our knowledge of the greenhouse effect is discussed and the limitations of existing atmospheric models that are used to predict changes in the greenhouse effect are pointed out. In particular the possibly dominant effect of the cloud/radiation feedback process is mentioned. Current models predict about a 2 K rise in surface temperature for a doubling of the atmospheric CO_2 concentration and a 6 K rise if the CO_2 increased by a factor of 8 as is possible in the future if the increase in the burning of fossil fuels continues. Possible greenhouse effects due to the release of chlorofluoromethanes (c.f.ms) and aerosols into the atmosphere from man-made sources are small. Analysis of the surface temperature changes which have occurred in the Northern Hemisphere during the last 100 years are consistent with most of the changes being due to increased CO_2 and decreased stratospheric aerosol.

1. The greenhouse effect

In the absence of clouds the Earth's atmosphere is largely transparent to solar radiation but mainly opaque to infrared radiation (figure 1). Because much of the radiation in the infrared emitted by the Earth's surface is trapped by the atmosphere, the temperature of the Earth's surface is substantially raised. This mechanism is known as the greenhouse effect. The name arises because the same mechanism occurs in glass greenhouses, glass being largely transparent to solar radiation while absorbing completely radiation in the infrared beyond about 3 μm in wavelength. Although the more important function of a greenhouse is to reduce the circulation of air, rather than to trap the infrared radiation (as is for instance exemplified by the effectiveness of greenhouses made of polythene which is transparent in the infrared as well as the visible), the name greenhouse effect continues to be used.

The greenhouse effect can be enhanced if the concentration of constituents which absorb in the infrared is increased. The constituents that I shall discuss in this paper are (1) carbon dioxide, (2) minor constituents such as the chlorofluoromethanes (c.f.ms) which possess absorption bands in the 10–12 μm atmospheric window where there is relatively little absorption by other constituents, and (3) aerosol.

2. Carbon dioxide

Carbon dioxide is the main absorber of atmospheric radiation in the 13–18 μm wavelength region – a region near the peak of energy distribution for black bodies near terrestrial temperatures (figure 1). Because of the burning of fossil fuels the concentration of carbon dioxide in the atmosphere has been increasing and will increase more rapidly in the future as the world consumption of fossil fuels accelerates. The mean concentration of CO_2 in the atmosphere measured at Mauna Loa, Hawaii, is currently (1977) 329 parts/10^6 by volume and is rising at about 0.9 parts/10^6 per year. (Peterson *et al.* 1977.) Before industrialization its concentration is estimated as having been around 280 parts/10^6, which implies that about half of the carbon dioxide resulting from normal activities has remained in the atmosphere (*Inadvertent climate modification*

1971). By the year 2000 the CO_2 concentration is expected to rise to 380 parts/10^6, and on the assumption that most of the known world's supply of fossil fuels are burnt by the end of the twenty-second century, the amount of carbon dioxide in the atmosphere could rise by a factor of between 4 and 8 above the present amount (National Research Council 1977). This latter prediction is very uncertain because of our current lack of knowledge concerning the rate at which CO_2 is exchanged between atmosphere and ocean and because we have only very poor information about the changes which are occurring or which may occur in the carbon content of the biosphere (Bolin 1975).

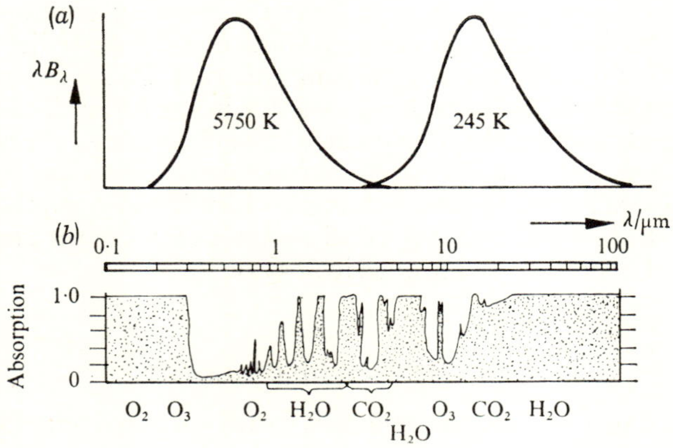

FIGURE 1. (a) Curves of black-body energy B_λ at wavelength λ for 5750 K (approximating to the Sun's temperature) and 245 K (approximating to the atmosphere's mean temperature). The curves have been drawn of equal areas since integrated over the Earth's surface and all angles the solar and terrestrial fluxes are equal. (b) Absorption by atmospheric gases for a clear vertical column of atmosphere. The positions of the absorption bands of the main constituents are marked. (From Houghton 1977.)

Since Plass's (1956) modelling study, a number of models have been developed for estimating the effect of increasing atmospheric carbon dioxide on surface temperature. The simplest of these involve calculations of the upward and downward radiation fluxes under some average atmospheric conditions from which a change in equilibrium surface temperature can be deduced following a change in carbon dioxide. More elaborate calculations involve fully developed three-dimensional models. One of the most thorough calculations to date is that by Manabe & Wetherald (1975). The model they employed was fully three-dimensional, and includes horizontal transports, convective adjustment, atmospheric water vapour dependent on surface conditions, and climatological cloud cover. The distribution of temperature change found from their model as a result of doubling the atmospheric CO_2 concentration is shown in figure 2; the average temperature change at the surface is about 2 K. Notice the much larger temperature change that is predicted near the pole, arising from what is known as ice–albedo feedback, i.e. if increased temperature removed some ice from the surface the surface albedo is reduced leading to further heating and increased surface temperature.

This figure of 2 K temperature rise resulting from a doubling of the CO_2 as found by Manabe & Wetherald (1975) is supported by the results of other similar models (see, for example, Schneider (1975) and is the one which is most often quoted. However, it must be pointed out that although these models include many of the relevant physical processes, there are some very fundamental feedback mechanisms (see figure 3) which are not included and for which there are, as yet, no adequate means available for their inclusion. The most important of these is the effect of varying cloud cover. In Manabe & Wetherald's model the cloud amount is not allowed to vary interactively with other parameters, whereas in the real atmosphere it depends on the detailed circulation patterns and also may be expected to vary with surface temperature and water vapour concentration. In considering how the surface temperature would vary with changing cloud cover, two competing effects have to be taken into account for an increase in cloud cover,

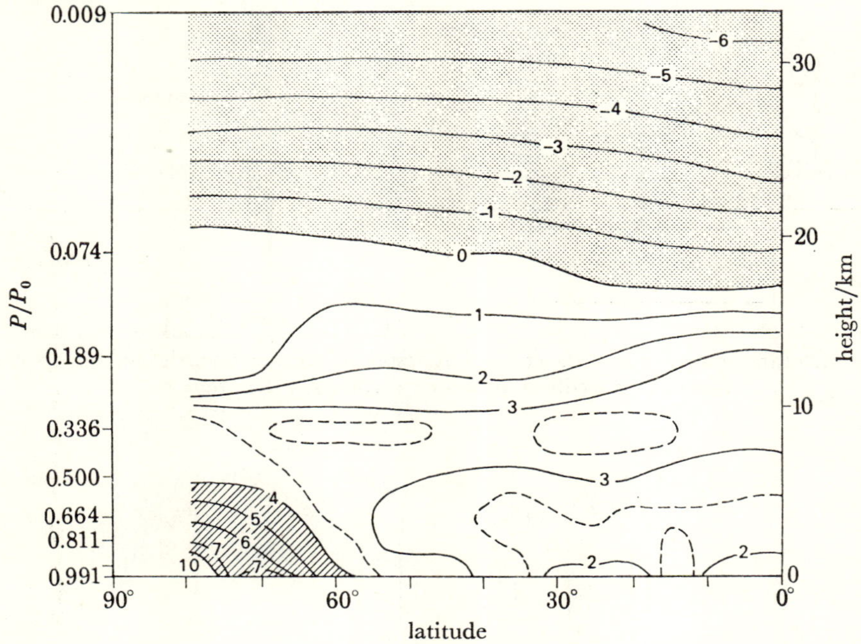

FIGURE 2. Distribution of temperature change (kelvins) resulting from a doubling of the CO_2 concentration as computed in a general circulation model by Manabe & Wetherald (1975).

namely the loss of solar radiation due to the increased albedo and the increased downward infrared flux at the surface. Cess (1976) suggests that the two effects approximately balance each other out, whereas Schneider (1972) finds the albedo effect dominates so that an 8 % increase in cloud cover would reduce the surface temperature on average by about 2 K. Schneider (1972) also points out the importance of cloud-top height (an increase of less than 0.5 km would raise the surface temperature by 2 K) and that the variation of the resultant effect with latitude is quite marked. Of particular importance is likely to be the effect on cloud cover near the pole where the ice–albedo feedback can be so effective. Better observations of the details of the Earth's cloud cover and its variations with space and time are urgently needed so that more adequate methods of incorporating cloud into numerical models can be devised. In the meantime, considerable uncertainty regarding the accuracy of estimates of the effect of increasing CO_2 such as those of Manabe & Wetherald must remain.

The above discussion has centred around the surface temperature changes which would occur with a doubling of the atmospheric CO_2 content. As the CO_2 amount is increased further the response of the surface temperature as estimated by current models is not linear with CO_2 amount (see, for example, Rasool & Schneider 1971) but more nearly logarithmic so that an increase to eight times the present CO_2 concentration would, in these models, result in an average surface temperature rise of around 6 K.

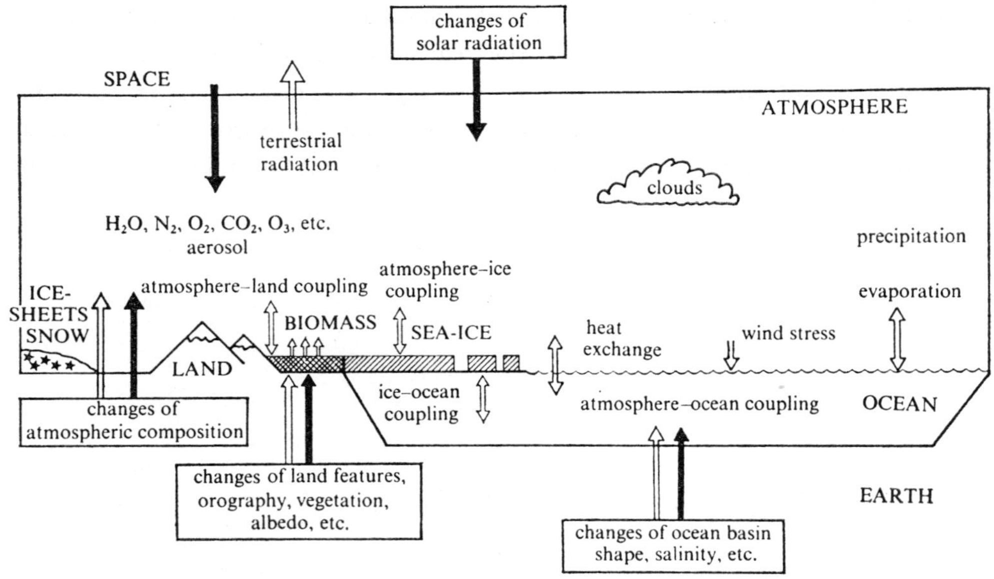

FIGURE 3. Schematic of the components of the coupled atmosphere–ocean–ice–earth climatic system illustrating some of the feedback processes. (From G.A.R.P. 1975.)

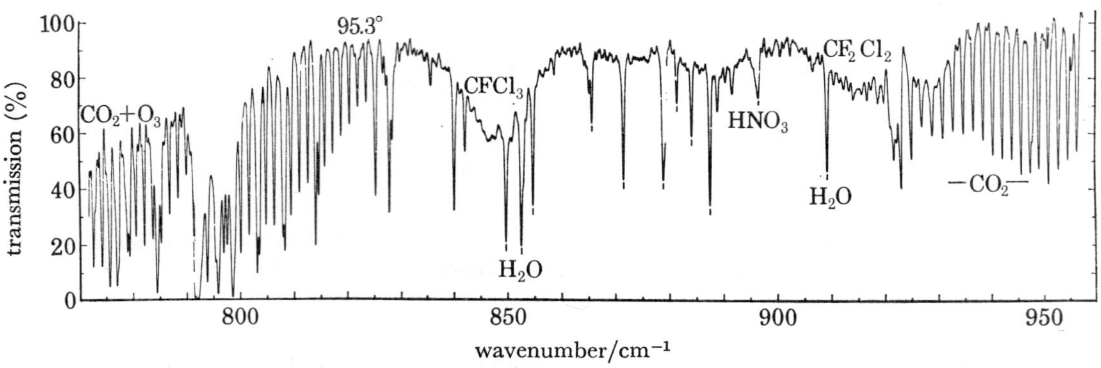

FIGURE 4. Atmospheric transmission in the atmospheric window region observed from a balloon-borne spectrometer from a height of 30 km along a limb path at 95.3° to the zenith by Williams *et al.* (1976). Notice the absorption bands of $CFCl_3$ and CF_2Cl_2. Resolution is 0.2 cm^{-1}.

3. The chlorofluoromethanes

Figure 4 is a record of atmospheric transmission in the atmospheric window region near 11 μm wavelength of a near horizontal path at an altitude of 30 km as measured from a balloon. Absorption bands due to $CFCl_3$ and CF_2Cl_2 can be clearly seen. Although for a vertical path of atmosphere the maximum absorption in these bands is small – less than 0.5 % – because they are in the window regions and near the peak of the energy spectrum for a black-body at terrestrial temperatures, it is possible that the effect on radiative transfer could be significant. Using a simple model, Ramanathan (1975) has calculated that if c.f.ms continue to be released at 1973 rates so that by the year 2000 the tropospheric concentrations were about 0.3 parts/10^6 for $CFCl_3$ and 0.6 parts/10^6 for CF_2Cl_2 (National Research Council 1976), the surface temperature rise would be about 0.2 K which is about half the effect which would arise from increased CO_2 during the same period. This model calculation is, of course, subject to the same assumptions and limitations as those for CO_2 mentioned in § 2.

4. Aerosols

Particulate matter, mainly originating from the surface, is present in the troposphere where its lifetime is quite short (typically a few days). Tropospheric aerosol plays an important part in determining the heat budget of the surface and atmosphere over large continental areas but may not be very significant so far as changes in the global climate are concerned.

Stratospheric aerosol, on the other hand, has a much longer lifetime (several years) and is spread around the globe so that climate on a global scale may be affected. There are well-defined aerosol layers present near 20 km altitude consisting largely of sulphate particles which are formed as a result of chemical processes in the stratosphere (Toon & Pollack 1973). Other sources of stratospheric aerosol are volcanoes (which may in addition produce the SO_2 for the sulphate chemistry) and debris from meteor showers.

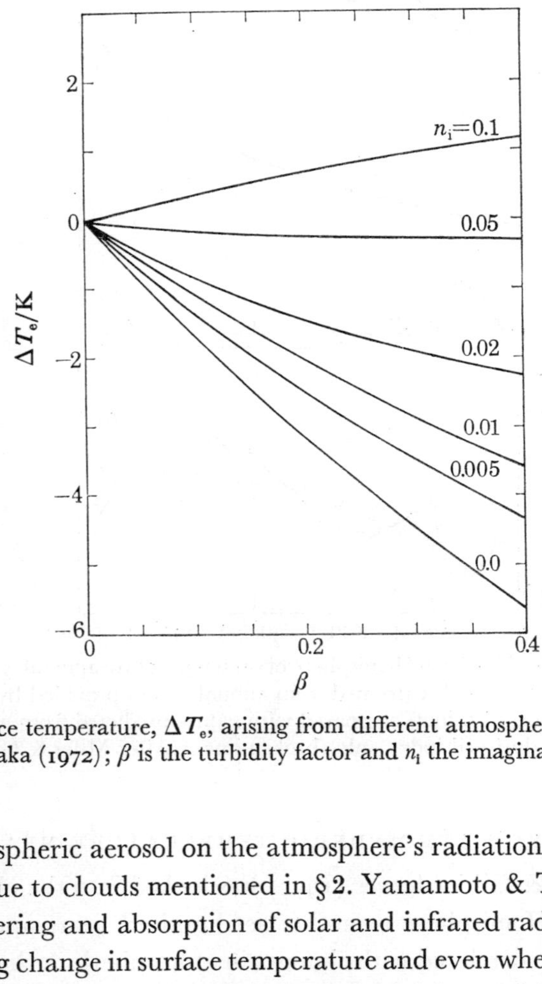

FIGURE 5. Change in surface temperature, ΔT_e, arising from different atmospheric aerosol contents as calculated by Yamamoto & Tanaka (1972); β is the turbidity factor and n_i the imaginary part of the refractive index of the particles.

The effects of stratospheric aerosol on the atmosphere's radiation budget arise from the same two reasons as those due to clouds mentioned in § 2. Yamamoto & Tanaka (1972) have made a detailed study of scattering and absorption of solar and infrared radiation by a layer of aerosol. The size of the resulting change in surface temperature and even whether it is positive or negative depends on the imaginary part n_i of the refractive index of the particles (figure 5). Typical values of n_i for stratospheric particles are 0.01 or less (Toon & Pollack 1976) so that cooling rather than heating can be expected to result from an increase in optical thickness of the stratospheric aerosol layer. By way of illustration, a situation of increased aerosol following the eruption of the volcano Fuego in October 1974 can be considered; it has been studied in some detail by Russell & Hake (1977). A typical optical thickness for the mid visible part of the spectrum of the stratospheric aerosol in a vertical path under unperturbed conditions is 0.005. After the Fuego eruption, the optical thickness rose to about 0.03. For such a layer a model of Harshvardhan & Cess (1976) predicts a cooling of the surface of about 0.8 K. The evidence that changes of surface temperature occur after volcanic eruptions has been reviewed by a number of authors (see, for example, Lamb 1970). Can we in addition expect changes in stratospheric aerosol due to man's activities? Baldwin *et al.* (1976) considered the possibility of a significant increase in stratospheric particles from high-flying aircraft and concluded that even if very large fleets of aircraft flew at high altitude, the optical thickness of the aerosols that they might produce would be several orders of magnitude less than for those from the Fuego volcano just discussed.

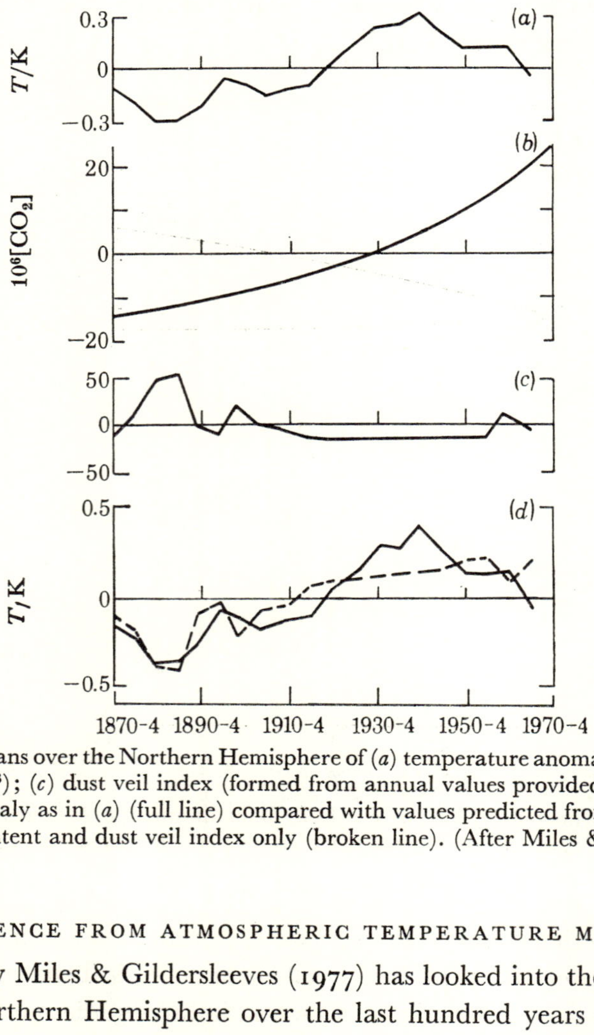

FIGURE 6. Five-year means over the Northern Hemisphere of (*a*) temperature anomaly; (*b*) CO_2 content (difference from 305 parts/10^6); (*c*) dust veil index (formed from annual values provided by Lamb (1970)); (*d*) actual temperature anomaly as in (*a*) (full line) compared with values predicted from multiple regression equation including CO_2 content and dust veil index only (broken line). (After Miles & Gildersleeves 1977.)

5. Evidence from atmospheric temperature measurements

A recent study by Miles & Gildersleeves (1977) has looked into the changes in average temperature of the Northern Hemisphere over the last hundred years and investigated to what extent these changes can be correlated with changes in atmospheric carbon dioxide and volcanic dust. Figure 6, taken from their paper, shows the basic data they employed. Their analysis shows that 65 % of the variance of hemispheric temperature can be explained by a multiple regression equation including only CO_2 concentration and volcanic dust veil index (figure 6). On the basis of this analysis, 0.4 K of the warming of *ca.* 0.6 K up to 1940 resulted from CO_2 (0.1 K) and from clearance of the dust veil (0.3 K). Miles & Gildersleeves's analysis further attempted, by including the ice index (figure 6) in their regression equations, to estimate the magnitude of the feedback arising from the presence of arctic ice. They conclude that of the 0.4 K warming mentioned above, approximately one-third is a result of ice feedback.

If the value of 0.1 K for warming from CO_2 before 1940 is extrapolated to what would happen if CO_2 increased by a factor of 2, a warming of about 1.5 K results, a very similar value to that found from the models reported in § 2. It must, however, be emphasized very strongly that the existence of correlations does not in any sense prove causal connections. However, as Miles & Gildersleeves point out, they do put some limit to speculation.

References (Houghton)

Baldwin, B., Pollack, J. B., Summers, A., Toon, O. B., Sagan, C. & Van Camp, W. 1976 *Nature, Lond.* **263**, 551–555.

Bolin, B. 1975 In *The physical basis of climate and climate modelling*. G.A.R.P. Publication Series, no. 16.

Cess, R. D. 1976 *J. atmos. Sci.* **33**, 1831–1843.

G.A.R.P. 1975 *The physical basis of climate modelling*. G.A.R.P. Publications Series, no. 16.

Harshvardhan & Cess, R. D. 1976 *Tellus* **28**, 1–10.

Houghton, J. T. 1977 *The physics of atmospheres*. Cambridge University Press.

Inadvertent climate modification 1977 Report of the study of man's impact on climate. M.I.T. Press.

Lamb, H. H. 1970 *Phil. Trans. R. Soc. Lond.* A **266**, 425–533.

Manabe, S. & Wetherald, R. T. 1975 *J. atmos. Sci.* **32**, 3–15.

Miles, M. K. & Gildersleeves, P. B. 1977 *Met. Mag.* **106**, 314–322.

National Research Council 1977 *Energy and climate* (*Studies in geophysics*). Washington, D.C.: National Academy of Sciences.

National Research Council 1976 *Halocarbons: environmental reports of chlorofluoromethane release*. Washington. National Academy of Sciences.

Peterson, J. T., Komhyr, W. D., Harris, T. B. & Chin, J. F. S. 1977 *Geophys. Res. Lett.* **4**, 354–356.

Plass, G. N. 1956 *Tellus* **8**, 140–154.

Ramanathan, V. 1975 *Science, N.Y.* **190**, 50–52.

Rasool, S. I. & Schneider, S. H. 1971 *Science, N.Y.* **173**, 138–141.

Russell, P. B. & Hake, R. D. Jr 1977 *J. atmos. Sci.* **34**, 163–177.

Schneider, S. H. 1972 *J. atmos. Sci.* **29**, 1413–1422.

Schneider, S. H. 1975 *J. atmos. Sci.* **32**, 2060–2066.

Toon, O. B. & Pollack, J. B. 1973 *J. geophys. Res.* **78**, 7051–7059.

Toon, O. B. & Pollack, J. B. 1976 *J. app. Met.* **15**, pp. 225–246.

Williams, W. J., Kosters, J. J., Goldman, A. & Murcray, D. G. 1976 *Geophys. Res. Lett.* **3**, 379–382.

Yamamoto, G. & Tanaka, M. 1972 *J. atmos. Sci.* **29**, 1405–1412.

Phil. Trans. R. Soc. Lond. A. **290**, 523–542 (1979)
Printed in Great Britain

The dispersion of sulphur pollutants over western Europe

By F. B. Smith and R. D. Hunt
Meteorological Office (Met. O. 14), *London Road, Bracknell RG*12 2*SZ, U.K.*

The various problems associated with the dispersion, advection and deposition of sulphur pollutants emitted into the atmosphere are reviewed in the light of experience gained during several long-range transport projects, in particular the O.E.C.D. project recently completed over western Europe.

Modelling and budgeting the exchange of pollution between countries is briefly discussed and it is noted that although 'statistical' modelling may be satisfactory when considering long-term effects, the episodic nature of deposition and the complex meteorological flows often associated with these abnormally high depositions suggests that smaller-scale and detailed modelling is required if these situations are to be properly understood and forecast.

Introduction

Sulphur is emitted from all fossil-fuel burning plants into the atmosphere, mainly as sulphur dioxide although small concentrations of sulphate may also be present. Although some of this sulphur is deposited locally much of it is diffused into the atmosphere and carried by the winds over considerable distances. Eventually the sulphur is either dry-deposited or is removed by precipitation to the ground. Typically over Europe between 0.1 and 10 g S/m^2 are deposited each year by dry deposition, the larger values occurring near the major source regions, and between 0.2 and 2 g S/m^2 by wet deposition. While the larger wet depositions also occur near the major source regions they also tend to occur in areas of high rainfall induced by mountain ranges, as in southern Norway. It is often in these very wet areas that the soil is thin and acidic, and the sulphate coming down in the rain is not readily neutralized and flows into the streams and lakes when fish kill can result. The extent and nature of the damage as it is appreciated at the present time has been summarized in a report edited by Braekke (1976).

A major international study of the long-range transport of sulphur pollutants on a scale of 1000 km and more was initiated in 1971 by the Organisation for Economic Cooperation and Development (O.E.C.D.) and the final report is now available (O.E.C.D. 1977). The study initiated extensive measurements of air concentrations and concentrations-in-precipitation of various pollutants, principally sulphur dioxide and sulphates. Ground-level stations made 24 h average measurements; aircraft made occasional and shorter-period measurements. Mathematical–physical models were developed to simulate the source–dispersion–sink processes taking place in the atmosphere. These models attempted to provide information on the amount of pollution reaching each part of the chosen area and how this could be partitioned between local sources and more distance sources in each of the other countries of Europe. Essential to the study was the determination of an acceptable emission inventory for the study area and a means of calculating back-trajectories of the air carrying the pollution.

This paper relates the various factors influencing the transport of sulphur more or less in sequence from its emission, initial dispersion through the mixing layer, advection by the wind, dry deposition, chemical change and wet deposition in precipitation. Finally the modelling is briefly discussed followed by some of the budgeting results and a look at the potential importance of very heavy episodic depositions.

As far as possible, duplication of detailed results described in the O.E.C.D. report has been avoided. Nevertheless a fairly complete summary of the meteorological problems involved has been attempted, and, when relevant, specific contributions arising from the Meteorological Office's own work in the O.E.C.D. study have been included.

In general it may be concluded that relatively simple models such as those used in the O.E.C.D. study are capable of giving reasonable estimates of one country's annual contribution to another's total sulphur deposition, but that on a day-to-day basis, and particularly in complex episodic situations, more sophisticated modelling would be essential if, for example, sporadic emission control were ever to be considered feasible.

Emissions of sulphur dioxide

Most fossil fuels contain sulphur, a large proportion of which will escape in the waste gases to the atmosphere.

Typical sulphur contents as quoted by Weatherley (1977) from N.C.B., C.E.G.B. and *Petroleum Review*, April 1975, are roughly:

coal	1.2–1.5 % S;
fuel oil	2.5–3 % S;
other petroleum fuels	0.04–0.7 % S.

The number of sources are of course extremely numerous; they vary with time in location and strength and so it is virtually impossible to give precise emission figures over a short period. Annual figures over grid squares with sides of the order of 100 km can be estimated rather more accurately, especially in the United Kingdom where statistical records of fuel consumption are maintained. Nevertheless, uncertainties do remain and the annual emission for the whole of the U.K. is probably only accurate to within less than 10% and for some other European countries to within only about 30%.

It must follow that in studying air concentrations of sulphur affecting a region at a given time, one of the biggest uncertainties must be the distribution of upwind emissions.

One of the first priorities at the beginning of the O.E.C.D. study was to establish the best emission fields that could be obtained. The Meteorological Office made an analysis of the emissions within the U.K. (Smith & Jeffrey 1975) with the use of a considerable number of data for 1970. Figure 1, taken from their paper, shows emissions of sulphur dioxide in thousands of tonnes for the year within 20 × 20 km Ordnance Survey squares covering the U.K.

Weatherley (1977) has shown that emission values have tended to decrease since 1970, although not entirely monotonically, and in 1974 were some 12 % less at about 5.4 Mt for the whole of the U.K. This decrease reflects the growing use of natural gas from the North Sea. Emission maps therefore have this unfortunate tendency to get rather out of date.

The Central Coordinating Unit of the O.E.C.D. study, the Norwegian Institute of Air Research (N.I.L.U.), compiled such emission data as was available to provide a similar emission map for Europe but on a 127 × 127 km square grid. In areas where no emission values were available a necessarily subjective assessment had to be made based on population figures and a rough idea of industrialization.

In western European countries, excluding Norway and Switzerland where fossil fuels are largely replaced by water power in the generation of electricity, emissions are roughly proportional to population at the rate of approximately 100 kg SO_2 released into the atmosphere per year for each person.

Thus the U.K. emissions worked out on this basis are $54 \times 10^6 \times 100$ kg SO_2, i.e. 5.4 Mt, in rather fortuitous agreement with the 1974 direct estimate. France, F.R.G. and Holland claim to have a lower rate, nearer 60 kg SO_2 a^{-1} per head, although this could either be simply a reflexion of rather less exact data or the use of substantially more natural gas or low-sulphur fuel oil.

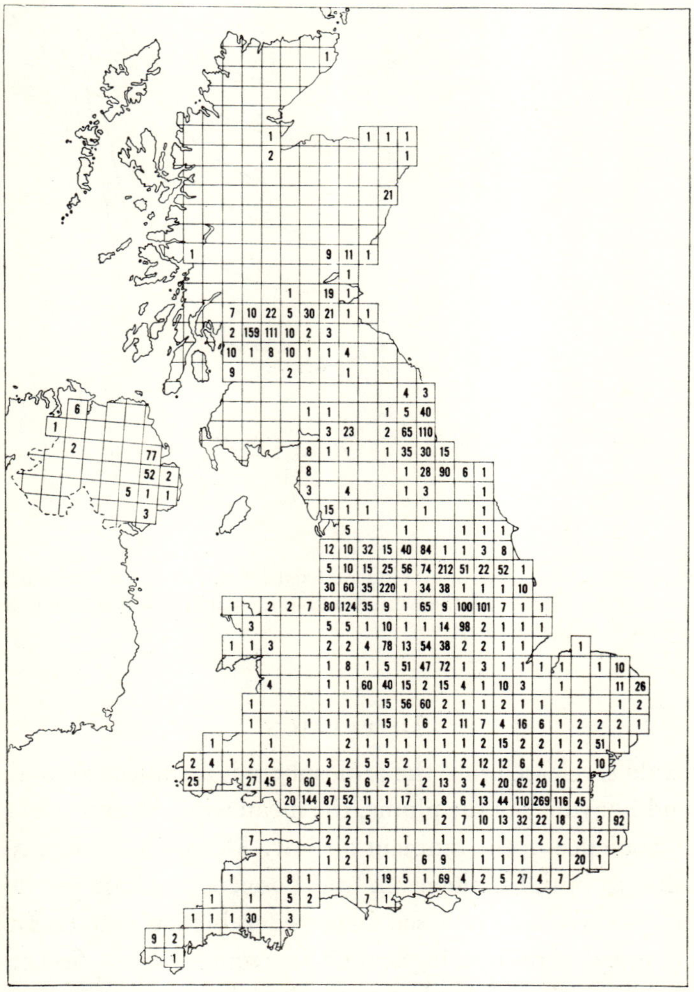

FIGURE 1. United Kingdom sulphur dioxide emissions for 1970 in thousands of tonnes per year. Each square represents an Ordnance Survey 20×20 km region.

One factor concerning emissions relevant to the long-range transport problem is the height at which the sulphur is emitted. As will be shown in the next section, this has a small but significant effect on the fraction of sulphur dioxide deposited on the ground before the plume is thoroughly mixed through the boundary layer.

In the U.K. over the 5-year period 1969–74 the percentage of the total emission attributed to high-level sources (power stations and oil refineries) has risen steadily from 47 to 58 %, whereas the percentage from medium-level sources (other industries) has fallen from 35 to 29 %, and from low-level sources (domestic, transport, etc.) from 18 to 13 % (see Weatherley 1977). Overall these trends must increase the amount being advected to distant regions, thereby slightly increasing the depositions and resulting effects, if any, but at the same time decreasing relatively much more the close-to-source air concentrations and depositions resulting in very noticeable improvements in the health of local inhabitants and reductions in damage to buildings and plant life.

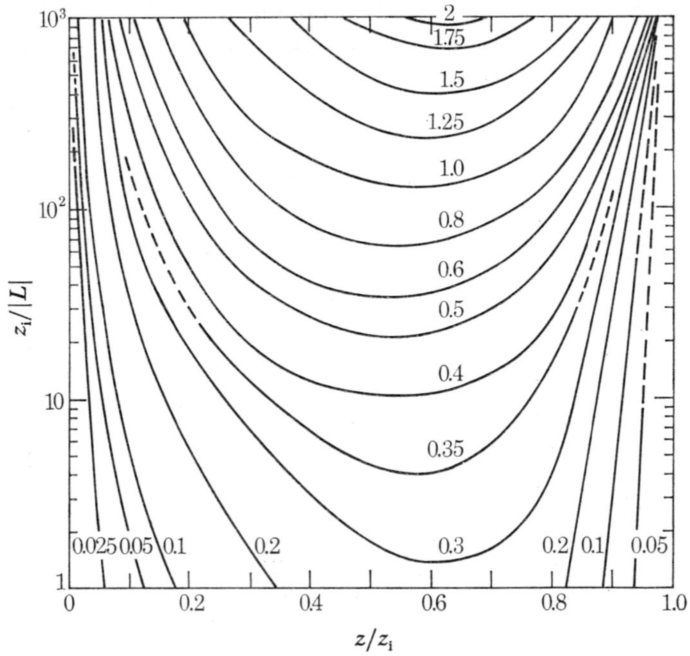

FIGURE 2. Contours of $K(z)/u_*z_i$ as a function of normalized height, z/z_i, and stability, $z_i/|L|$, where z_i is the mixing depth, u_* the friction velocity, L the Monin–Obukhov length scale, based on data from the Minnesota field experiment (Izumi & Caughey 1976).

DISPERSION

Once sulphur dioxide is released into the atmosphere it is subject to motions arising from the initial momentum and buoyancy of the plume, the mean wind, the wake of the chimney and the whole spectrum of turbulence in the ambient airflow. Many formulae have been presented to predict plume rise due to initial momentum and buoyancy. Those due to Briggs (1969) and Moore (1974) are best known and are based on considerable sets of data. In general the result is to replace the real chimney height by a higher effective source height. Sometimes the plume may penetrate through the top of the mixing layer into the relatively non-turbulent air above, when the sulphur can be transported over considerable distances effectively insulated from the sink at the ground, until ultimately by one process or another the plume is reincorporated into the mixing layer or the sulphur is washed out in precipitation.

For most low-level and medium-level sources, however, the plume rise is fairly small and in day-time at least the plume diffuses to the ground well before it diffuses to the top of the mixing layer. In these situations the concentration at ground level rises well above that predicted by the so-called box model, where the pollution is assumed to be uniformly mixed through the whole of the mixing layer. Dry deposition to the ground will be correspondingly high until the concentration falls to the box model value in a time or distance which depends very much on the stability and depth of the layer, z_i. The stability, μ_i, may be represented by a ratio of two lengths: z_i itself and the Monin–Obukhov length-scale, L, i.e.:

$$\mu_i = z_i/|L|.$$

According to an analysis by Smith (1977) of basic boundary-layer data taken from the Wangara experiment (Clarke *et al.* 1971) and from the Minnesota field experiment (Izumi & Caughey 1976) the diffusivity $K(z)$ can be expressed in similarity terms as shown in figure 2. The implication of this result is that the surface concentration will reach within 4 % of its box model value after a distance of travel x given in day-time 'unstable' conditions by table 1, provided the surface Rossby number G/fz_0 is close to its normal land value of 10^6. Thus if the geostrophic wind G is $10\,\mathrm{m\,s^{-1}}$ the coriolis parameter f is $10^{-4}\,\mathrm{s^{-1}}$, the surface roughness $z_0 = 0.1$ m, the upward surface sensible heat flux H is $200\,\mathrm{W\,m^{-2}}$ and z_i is 1200 m, then $\mu_i = 40$ and $x \approx 10$ km.

TABLE 1

μ_i	1	4	10	40	100	400	1000
x/z_i	25	18	13.5	8.5	4.8	2.5	1

If, however, the geostrophic wind fell to $1\,\mathrm{m\,s^{-1}}$ the value of z_i would not change appreciably but G/fz_0 would decrease to 10^{-5}, μ_i would increase to about 1000 and x decrease to about 1 km. In terms of time and also in terms of the fraction of SO_2 deposited, very little change would be observed. The deposition in the first case, however, would be spread over roughly ten times the area of that in the second.

Most low- and medium-level sources are situated in urban areas where two factors affect the vertical dispersion and related deposition. First, an urban area is a limited region of increased surface roughness. Secondly, it is an area where the upward surface sensible heat flux is significantly greater than in the surrounding countryside (frequently by as much as $70\,\mathrm{W\,m^{-2}}$) because of man-made heat sources and because less of the incoming natural radiation is used in evaporating surface moisture: a city is a relatively well drained and arid area. Both these effects tend to increase vertical dispersion and tend to militate against the formation of stable layers which would trap low-level emissions and give greatly enhanced depositions. Limited convergence of air in the area at low levels may also be induced, although these should not greatly affect the depositions.

Altogether then, accurate determination of the enhanced deposition close to the source over and above that given by the box model is fraught with difficulty but fairly simple calculations based on, for example, the Hanna–Gifford (1973) model of urban diffusion suggests that roughly an extra 10–20 % of the emitted SO_2 is likely to be deposited (Smith & Hunt 1977) within the first 50 km or so.

At night the air over the urban area may be nearly neutrally stratified whereas elsewhere it is stable. The urban plume may then be advected over the stable and slow-moving surface air in the countryside as it moves away and will be protected from further deposition until fumigation occurs in the following morning. In less-concentrated urban areas the air may be everywhere stable on most nights, the low-level plumes may be retained close to the ground and virtually all the sulphur dioxide will be lost by surface absorption.

Plumes disperse in the horizontal as well as the vertical. Unlike vertical turbulence where the range of eddy scales is limited by the depth of the mixing layer, with horizontal turbulence a fairly continuous spectrum exists from the very smallest eddies, through meso-scale eddies (range

1–100 km) up to synoptic-scale eddies of the size of anticyclones and depressions. The across-wind width of the plume is therefore very dependent on the time over which it is sampled: the longer the time, the broader the plume. Since on the one hand episodic deposition levels usually persist for several hours, and on the other, ground-level measurements are effective averages over 12 or 24 h, the width of the plume (or conversely the upwind source band affecting the receptor point) is governed principally by eddies in the synoptic range, and eddies in the micro scale and perhaps also the meso-scale are virtually unimportant. Fortunately variations in the synoptic pattern can be traced reasonably successfully from the network of meteorological measuring stations, and with somewhat less success into the immediate future using numerical weather forecasting models.

Advection

Although trajectory analysis has been carried out by meteorologists for a very long time to track air masses, important in forecasting future weather, the precision with which this can be done is not particularly high. The main reason for this (and this applies particularly to pollution released in the lower part of the atmosphere) is that wind varies in strength and direction with height. In ideal circumstances, when the pressure gradient is virtually constant in time and space and the underlying surface does not change character on scales of 100 km or more, boundary-layer theory predicts how wind speed will decrease as the ground surface is approached, and how wind direction will back away from the geostrophic direction until close to the ground the wind will have a component down the pressure gradient towards low pressure with magnitude depending on surface roughness, the geostrophic wind, latitude and air stability. Typically the surface wind will be backed some 10° over the sea, 20° over land during the day and 30° or more at night.

In more realistic situations, air parcels are normally undergoing acceleration or retardation which affects the balance of forces and the resulting flow vector. For example, boundary-layer air diverging slowly out of a slack anticyclonic region into a tighter low pressure system will be accelerating and will gain some of the required energy by increasing the angle of backing beyond the equilibrium values quoted above.

Mountains, sea breezes, marked temperature gradients, baroclinic effects and so on, can all cause variations of wind with height that are significantly different from the 'ideal'.

However, pollution in a well-mixed layer by definition is not advected by winds at one level alone but by all the winds in the layer as each element diffuses up and down by vertical turbulence. The stronger the vertical mixing, the more rapid is this sampling process and the more each and every element moves with the *average* velocity. Momentum is also efficiently exchanged in these circumstances and the layer cannot support large velocity shears anyway except very close to the ground. If the vertical mixing is relatively slight, the boundary layer (or mixing layer) is likely to be correspondingly shallow and again shears are not likely to be too important. Wind shear is in fact only likely to be important if either the boundary layer has shrunk leaving some of the pollution at levels now stably stratified where shear can play its full rôle, or if the boundary layer is still fairly deep but in a state of decreasing turbulent energy as in the latter part of the day.

Actual measurements of plumes over distance scales of the order of hundreds or even thousands of kilometres within the boundary layer are few and far between. What evidence exists suggests that trajectories in maritime conditions are usually easier to predict than, say, in winter–large continental conditions where strong stable stability results in very marked wind shears with height. Smith has analysed records of identifiable plumes originating in the U.K. and passing over Finland and has shown that in these westerly flows the average backing from geostrophic is 7° with an extremely small standard deviation of less than 4°. Measurements made by the Meteorological Research Flight aircraft during sulphur sampling flights in the boundary layer off the east coast of England confirm that in westerly flows the average backing is 10° in neutral or slightly unstable conditions with no significant reduction of speed below the geostrophic wind, whereas in slightly stable conditions the backing is nearer 20° with a 10 % reduction in speed (Smith & Jeffrey 1975). Proposed aircraft tracking of plumes carrying implanted tracers by the Central Electricity Research Laboratory and the Meteorological Office promise new and potentially very valuable data in the next few years on this topic.

Analysis of the behaviour of tetroons flying at a *constant* level in the boundary layer and being tracked by radar or by being returned to the launchers with details of their ultimate recovery point by members of the public, provides very useful information on the behaviour of low level winds. Pack (1977) has summarized many of these experiments mainly carried out in the U.S.A. by Angell, Peterson and others. Remembering that these are indicative of winds at just one level (about 100 m) and do not fully represent the behaviour of 'free' pollution plumes, the following are two of Pack's general conclusions:

(i) the backing in southerly flows is on average 20° from the surface geostrophic direction and about 40° for northerly flows;

(ii) the best trajectories were obtained using twice the 10 m wind speed veered by 10°, although even here the standard errors were not insignificant.

Pack also quotes radar-tracked tetroon experiments carried out by the Air Resources Laboratory at Las Vegas (Allen *et al.* 1967) in mountainous terrain. Using rather sparse wind data, Allen attempted to interpolate to estimate the winds affecting the tetroons. The effects of the terrain, etc., were however so great that the reconstructed trajectories were on average to the left of the real flight paths by roughly 20 % of the distance of travel, and the standard deviation was as much as 60 %.

Returning to pollution plumes, the best trajectories should be attained by using interpolated real measured winds, and the Heffter N.O.A.A. model (1975) provides an excellent basis for doing this. In many areas and especially over sea areas, real winds are not available in high enough spatial density or with great precision by current operational sondes. The use of the surface geostrophic wind is a very attractive alternative since it is based on relatively plentiful surface pressure observations made at 1 h or 3 h intervals. Sykes & Hatton (1976) has provided a sophisticated numerical interpolation scheme which uses these to predict trajectories. Such trajectories are of course subject to all the errors and uncertainties outlined earlier, even when mean backing and speed reductions are applied.

The Norwegian team at N.I.L.U. preferred to use 850 mbar† observed radiosonde winds to calculate their trajectories. Although these apply to a rather high level (typically 1500 m) which may often be above the boundary layer they have the advantage of being actual observed winds. Comparison of trajectories worked out by N.I.L.U. and by ourselves at the Meteorological Office with the use of modified surface geostrophic winds shows generally good agreement. The largest discrepancies occur in southerly winds when surface geostrophic winds permit substantial south to north advection, whereas the higher 850 mbar winds tend to curve the back-trajectories to

† 1 mbar = 10^2 Pa.

come in from the west. Looking at Norway as a receptor area this tends to decrease the contributions from source areas to the south and southeast in the N.I.L.U. results, to increase the contributions from countries to the southwest, principally the U.K. compared with the Meteorological Office model.

In summary, use of modified surface geostrophic winds, interpolated actual 850 mbar winds or interpolated mean boundary layer winds all have their particular advantages and disadvantages. More tracer studies are required to help decide which winds should be used taking into account availability, accuracy and frequency. Differences will almost certainly be apparent in single situations but it appears from the results of the various models that when averaged over a year or more, budgets of pollution transmission between countries are not very seriously affected by the choice.

TABLE 2

surface (moist climates)	v_g (mean values)
short grass	0.5
medium crops	0.7
forests	?1.0
calcareous soil	0.8
acid/peaty soil	0.5
typical mixed countryside	0.8
snow	0.1
water	0.7
cities	?0.7

DRY DEPOSITION

The gas, sulphur dioxide, is fairly readily absorbed by vegetation, soil and water. The rate of absorption is usually expressed as a velocity of deposition, v_g, which is defined as the ratio of the vertical downward flux to the surface to the air concentration. Since the latter is an increasing function with height, the velocity of deposition unfortunately has the property of being a function of height, and strictly also a function of stability. Some of the variability in quoted values of v_g no doubt arises from the failure to standardize to a given low height and a specified stability. In this respect the use of a surface resistance as proposed by Chamberlain (1966) has many advantages. However, v_g is still used, mainly because it has a simpler physical concept and because it is more readily incorporated into the basic conservation equations.

At the recent International Symposium on Sulphur in the Atmosphere (I.S.S.A.) held at Dubrovnik (1977) a very useful workshop summary on deposition was prepared, and several details in this section are quoted directly from it.

Table 2 summarizes some of the estimates of v_g in centimetres per second for sulphur dioxide. As mentioned in the next section, sulphur dioxide in the atmosphere undergoes gradual oxidation to sulphate if it is not dry-deposited first. The sulphate is usually in the form of a fine aerosol (0.1–1.0 μm) which also has a deposition rate to the surface. Most theories and measurements imply that v_g for sulphate is considerably less than that for sulphur dioxide and does not exceed 0.1 cm s^{-1}.

It has already been estimated that between 10 and 20 % of the emitted sulphur dioxide is deposited on average within the first few tens of kilometres while the plume is relatively concentrated near the ground. Thereafter the box model is reasonably applicable and the time-scale, τ, for the concentration to fall to $1/e$ of its original value is z_i/v_g when z_i is the mixing depth. If

$z_i = 1000$ m and $v_g \approx 0.01$ m s^{-1}, then $\tau \approx 10^5$ s or roughly 1 day. In reality, during one day, stabilization of the lower layers will occur insulating most of the sulphur dioxide from deposition to the ground. Consequently the actual time-scale is rather larger and lies generally between 1 and 2 days. As the I.S.S.A. summary aptly points out, this deposition is very important to general global concentrations; without it and without removal in precipitation the concentration would rise by some 70 μg m^{-3} per year.

The aircraft sampling flights over the North Sea described by Smith & Jeffrey (1975) were originated to compare total fluxes against upstream emissions and deduce losses to the intervening ground by dry deposition and the conversion rate of sulphur dioxide to sulphate. In 1971 the sampling system did not include sulphate and so nothing could be deduced directly about v_g. In 1973 this omission was rectified and the later results consistently pointed to a value of 0.8 cm s^{-1} over land and 0.7 cm s^{-1} over the sea, in full agreement with the values in table 2. The implication for the U.K. is that some 30 % of the emitted sulphur dioxide is dry deposited within our borders, that is roughly 1.7 Mt. It has been argued by F. Ross that part of this deposition is actually beneficial to agriculture in providing a source of sulphur essential for plant growth. In the main, agricultural areas of Britain the average air concentration is about 30 μg m^{-3} and the dry deposition rates range from some 50 to 200 kg S/ha each year.

Average rural concentrations, $\bar{C}$, in other countries can be roughly estimated from simple emission versus advection considerations, and provided the sulphur dioxide concentration largely originates within the country and not from neighbouring regions then the simple formula $\bar{C} = 90e/\sqrt{a}$, where e is the emission in megatonnes per year and a the area in thousands of square kilometres, gives an answer correct to within at least a factor of two. Dry depositions are then roughly $2\bar{C}$ in units of kilograms S per hectare per year.

Chemical changes

The oxidation of sulphur dioxide to sulphate is such a complex subject that it would be inappropriate here to give more than the briefest of outlines of our present knowledge and the experience we have gained in our own aircraft sampling flights.

In the O.E.C.D. project the main reason for studying the conversion to sulphate is that sulphate is the longer lasting species and is largely responsible for the long range transport. In many of the major American projects (M.I.S.T.T., S.U.R.E., M.A. P3. S.) the interest has centred round the possible health hazard associated with sulphate aerosol when in sufficiently high concentrations. In these projects large single plumes or dominant urban plumes have been tracked with aircraft and the conversion rates deduced by seeing how the sulphate concentration varies with downwind distance. The oxidation can take place by one or more of three general processes: by so-called homogeneous gas phase reactions depending on photooxidation, in aqueous droplets where the oxidation usually depends on metal catalysts (e.g. Fe, Mn) or ozone (O_3), or on surfaces. Ammonia plays a rather inadequately defined rôle in converting the sulphuric acid to ammonium sulphate.

The rate of conversion as implied by these plume studies varies considerably according to the internal chemistry of the plume, the relative humidity of the ambient air and the time of day (or incoming radiation). In the M.I.S.T.T. study Husar *et al.* (1977) found conversion rates as low as 0.1–0.5 % per hour at night and between 0.5 and 6 % during the day in the Labadie power plant plume.

In the Meteorological Office flights (Smith & Hunt 1977) the plumes were in general less intense and more diffuse. Nevertheless, at distances of 80–220 km downwind from the main upwind source areas, between 8 and 20 % of the original sulphur dioxide had been converted to sulphate. The interesting fact was that the conversion factor at these distances did not obviously correlate with either distance or time of travel but depended solely on relative humidity (see figure 3). The implication must be that the major part of the sulphate had been formed close to source where concentrations of sulphur dioxide and other relevant pollutants would be high, and that subsequent conversion once the plumes filled the mixing layer was sufficiently small to be masked by this initial conversion.

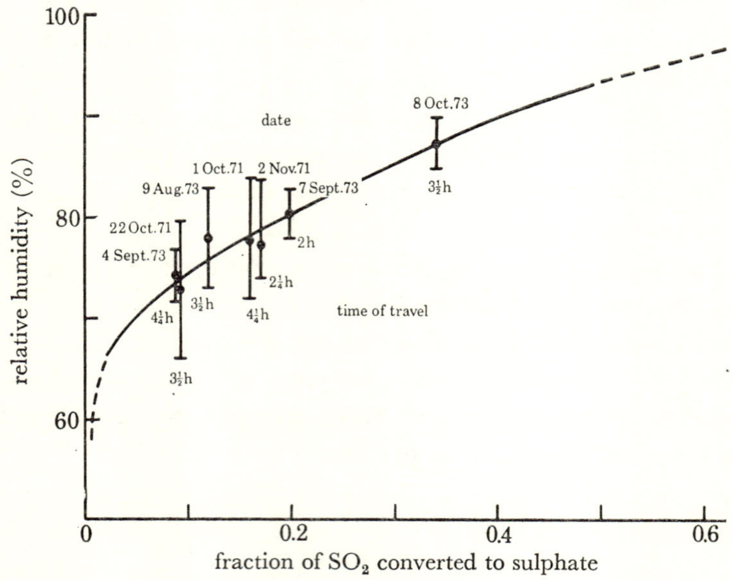

FIGURE 3. Implied fraction of the emitted SO_2 converted to sulphate as measured along the flight track off the East Coast of the U.K., as a function of relative humidity.

In order to try to determine the conversion rate once dispersion was complete through the mixing layer, five sampling flights were made during 1974–5 along the eastern North Sea either just off SW Norway or off west Denmark. Originally it has been intended to sample the same air late in the afternoon off the U.K. east coast and early the next morning on the other side of the North Sea, but this proved too difficult, especially without the use of tracers. As a result only on one flight, that of 22 August 1974, can an estimate of the conversion rate be made with any confidence. On this occasion there were two layers: a lower layer with an inversion extending up to about 260 m in which nearly all the sulphur dioxide had been deposited or converted to sulphate, and an upper layer in which the loss of sulphur dioxide was assumed to be due to conversion to sulphate only. Assuming that both layers had had identical sulphur dioxide and sulphate concentrations on leaving the U.K. east coast 18 h earlier, time constants for the exponential decay of sulphur dioxide by deposition and by conversion to sulphate were 7.3 and 99 h respectively. The latter is equivalent to a conversion rate of 1 % h^{-1}, which, while consistent with the American experience, confirms that with the present levels of precision in the aircraft system, variations of conversion with other parameters would be very difficult to detect with any certainty.

Wet deposition

The second major sink of sulphur in the atmosphere is removal through precipitation events. Typically in western European air, parcels will be advected into a rain (or snow) belt in a matter of a few days; the average time-scale is of the order of 3 days. Whenever significant precipitation is occurring, low-level convergence is almost always present to feed fresh water vapour into the cloud systems; otherwise the clouds would soon dry out. On the assumption that in a moderate rain situation the liquid water content of the cloud is not changing very significantly with time, the following simple sums show that subcloud layer convergence is sufficient to maintain the rainfall and to explain the observed wet deposition rates of sulphur in western Europe.

We suppose that the air in the subcloud layer has a relative humidity of 90 %, and at a temperature of 10 °C this corresponds to a mixing ratio of 10 g water per kilogram of air (that is roughly 1 m^3 of air). Although our arguments are not restricted to any particular rate of rainfall it is useful to digress and look at the case of moderate rain. Moderate rain is defined as a rate between 1 and 4 mm rain per hour. If this rain were to be balanced by an input of water vapour from the sub-cloud layer into the cloud, these rates would require vertical velocities between 3 and 12 cm s^{-1}, very reasonable values and similar to those actually measured. In general, however, if the concentration of sulphur (as SO_2 or sulphate) in the mixing layer is taken as approximately 10 μg S/m^3 (as typically observed in aircraft sampling flights), then on the assumption that all the sulphur is removed by one process or another into the rain, then 1 mm of rain, which corresponds to 100 m^3 of mixing layer air, must carry 100 × 10 μg S (i.e. 1 mg S). This sulphur burden is equivalent to a concentration of 3 mg SO_4/l, which is in very good agreement with measured values over a large part of NW Europe. Moreover, in southern Norway where the annual rainfall has an average of about 1 m, the implied wet deposition is 1 g S a^{-1}, again in very good agreement with direct measurements. Generalizing this analysis, the depletion of sulphur out of the mixing layer, along a trajectory, due to removal in precipitation can therefore to a good approximation be related to the rate of rainfall and to the observed mixing ratios.

If Q is the sulphur still carried in the atmosphere arising from emissions over unit time from upwind sources, t is time in hours, h is the mixing depth in kilometres, m is the mixing ratio in grams per kilogram and r is the rainfall rate in millimetres per hour then $\mathrm{d}Q/\mathrm{d}t = -rQ/mh$, and the time-scale in hours of this process is given by $\tau_s = mh/r$.

At warm fronts the input of sulphur is mainly from the warm sector air during the general upsliding process within the sloping frontal zone. However, some boundary layer air may be drawn up from the colder air ahead of the front, although experience in trying to seed warm fronts with silver iodide released at ground level in order to monitor rainfall modification suggests that usually little air penetrates the frontal inversion into the main rain-producing cloud.

Some sulphur still in the boundary layer may get removed in rain by direct sulphur dioxide and sulphate scavenging as the rain drops fall through the polluted air. This is a complex process and is discussed in the workshop summaries of the I.S.S.A. conference (Husar (ed.) 1977). There are unproven suggestions that these processes are not normally dominant in wet deposition of sulphur.

Returning to the low level convergence process into rain areas, the modelling of these convergences is particularly difficult on a fairly course grid as used in all current numerical trajectory models. This is particularly true in large thunderstorm situations where considerable volumes of boundary layer air can be swept into these dynamic clouds and cleansed by washout. Upward velocities over regions below cloud 1–2 km across can be as large as $1–10\ m\ s^{-1}$, implying normal local horizontal convergence rates of 10^{-3} to $10^{-2}\ s^{-1}$. Such situations can lead to depositions of episodic proportions.

Another major problem in recording wet deposition and the associated loss from the trajectory is the general lack of a good areal coverage of rainfall. This is especially so in mountainous areas, where rainfall is strongly influenced by the topography and is spatially very variable, and also over sea areas like the North Sea where few or no data are available. Perhaps the biggest hope for the future in this problem is in the deployment of suitable weather radar. The successful use of radar in quantifying rainfall over areas 80 km in diameter has been described by Browning (1977) and Hill *et al.* (1977). In the next decade most of England and Wales will be covered in this way.

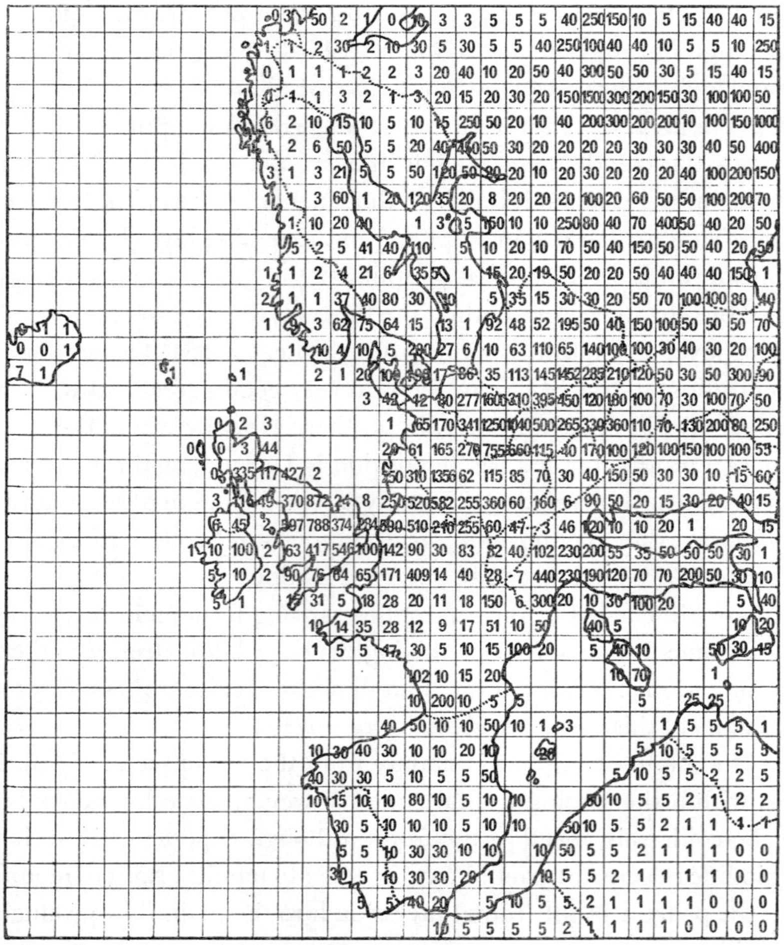

FIGURE 4. Sulphur dioxide emissions for 1972 in thousands of tonnes per year. The grid length is 127 km at 60° N.

MODELLING

Many attempts have been made at modelling some of the various processes outlined in the previous sections (see, for example, O.E.C.D. 1977; Scriven & Fisher 1975), both in order to produce sulphur deposition fields which agree satisfactorily with those produced from observations and, more particularly, to determine the fluxes of sulphur across national boundaries within

Europe. In the British Meteorological Office, a trajectory model has been developed (Smith & Johnson 1975) to calculate the relative contributions of different countries or regions in Europe to the total sulphur deposition in southern Norway. Back-trajectories were produced from a chosen point in Norway every twelve hours for the three year period 1972–4. The trajectories were based on surface geostrophic winds, modified in a simple fashion to make them more representative of winds in the boundary layer, and were taken back in time for at least 72 h or until precipitation was encountered. Each trajectory so produced was then related to a 127 km × 127 km emission grid for Europe (figure 4) and every grid square crossed by the trajectory was noted together with an indication of whether precipitation was occurring or not at the destination area in order to distinguish dry deposition from wet deposition occasions.

By assuming a constant mixing depth of 1 km and with the use of values of SO_2 deposition velocity and SO_2 to SO_4 transformation rate deduced from experiments discussed earlier, the amount of sulphur remaining in the air at the destination due to the emission in each crossed grid square could be determined and values for dry or wet deposition of sulphur were obtained. Although the model was essentially very simple, diurnal variations in emission and also diurnally varying amounts of local deposition and transformation at the source area were introduced. Some attempt was also made to include the effect on deposition of night-time stabilization. The contributions from all of the trajectories were then added to produce the results described in the next section.

Clearly this model, while producing useful results which agreed quite well with other models and tolerably well with observations, needs to be refined considerably to take into account many factors which are known to be of importance. Variations in mixing height and the corresponding wind shears encountered by the pollution should be introduced. Ways of estimating mixing height from readily available parameters have been developed (Smith 1975), but incorporating the effects of wind shear is difficult without introducing more than one level in the vertical. Introducing more detailed precipitation data is essential, with the use of any kind of measuring device available, especially over sea areas, while it will also be necessary to vary the dry deposition velocity according to the underlying surface.

One of the most important areas for improvement concerns the local depositions close to the sources. That not all of the sulphur emitted at a source is involved in long-range transmission has been discussed and allowances are made for this in the model. The term 'apparent emission' of a source can be used to describe that part of the total emission which is transported out of the source region. Exactly how much is deposited close to the source depends critically, as has been shown, on source height and on the mixing depth as well as on such mesoscale features as urban heat islands. Suitable ways of parameterizing the 'apparent emission' for each source at different times will be incorporated into future models.

Budgets

Before outlining some of the results produced by the model described above, some idea of the uncertainties involved in the analyses should be given. As has already been stated, the accuracy of the national emission totals is thought to be between about 10 and 25 % for most of the countries which contributed to the O.E.C.D. experiment. For other countries the figures are probably less accurate while the individual grid square values are less accurate still. The accuracy of trajectories is hard to verify. The use of different advecting winds (such as the 850 mbar wind which has been used in some models (O.E.C.D. 1977)) can lead to startlingly different trajectories on certain occasions and individual cases might be quite inaccurate. Nevertheless, systematic errors over a three year period could be relatively unimportant.

The wet deposition figures are greatly influenced by the simple manner precipitation is included in the model and the combined wet and dry deposition totals are probably only accurate to within about 40–50 %. Also, comparing results from the model with observations from a

Table 3. Average contributions by regions 1972–1974

region	distribution of wet deposition (%)	distribution of dry deposition (%)	distribution of total deposition (%)	total deposition mg SO_2 m^{-2} a^{-1}: annual average	total deposition mg SO_2 m^{-2} a^{-1}: per 100 kt emission
British Isles	34.4	22.2	30.7	895	15.0
Norway	10.9	32.2	17.4	507	323
G.D.R.	11.0	9.9	10.7	312	6.3
F.R.G.	11.4	7.2	10.1	295	7.3
Sweden	4.4	6.9	5.2	152	14.8
Holland	4.3	2.9	3.9	114	9.5
Poland	3.4	3.9	3.5	102	2.7
Denmark	3.3	3.7	3.4	99	26.2
France	4.1	1.9	3.4	99	4.1
Belg./Lux.	3.9	2.3	3.4	99	8.1
Czech.	3.5	2.7	3.2	93	2.9
U.S.S.R.	2.3	2.4	2.4	70	0.7
Italy	1.1	0.5	0.9	26	0.9
Finland	0.5	0.5	0.5	14	3.0
others	1.5	0.8	1.3	38	1.0
total	**100.0**	**100.0**	**100.0**	**2915**	

annual average wet deposition 2022 mg SO_2 m^{-2} a^{-1}
annual average dry deposition 893 mg SO_2 m^{-2} a^{-1}

Table 4. Deposition by regions for each year 1972–1974
(percentages of grand total for each year)

region	1972 total: 3698 mg m^{-2}		1973 total: 2231 mg m^{-2}		1974 total: 2823 mg m^{-2}	
	dry deposition	deposition in rain	dry deposition	deposition in rain	dry deposition	deposition in rain
British Isles	5.3	20.9	7.2	22.6	8.5	28.7
Norway	7.8	5.5	12.7	11.3	10.3	7.5
G.D.R.	3.9	11.3	2.5	5.1	2.3	4.9
F.R.G.	2.2	8.5	1.8	6.9	2.5	7.9
Sweden	1.9	3.1	2.1	3.5	2.4	2.7
Holland	0.9	3.4	1.0	3.2	0.9	2.4
Poland	1.1	2.9	0.8	2.0	1.6	1.8
Denmark	1.4	3.1	1.0	2.1	0.9	1.4
France	0.7	3.2	0.5	2.3	0.5	2.9
Belg./Lux.	0.7	2.9	0.8	2.6	0.6	2.4
Czech.	1.1	3.9	0.8	1.7	0.5	1.1
U.S.S.R.	0.5	1.0	0.5	3.3	1.2	1.1
Italy	0.2	1.0	0.1	0.0	0.1	1.0
Finland	0.1	0.1	0.1	0.3	0.3	0.6
others	0.2	1.2	0.2	1.0	0.3	0.7
total	**28.0**	**72.0**	**32.1**	**67.9**	**32.9**	**67.1**

number of measuring stations in southern Norway, it was apparent that the model tended to overestimate the amount of wet deposition and underestimate the dry deposition.

However, bearing the foregoing comments in mind, the results from the model are summarized in tables 3 and 4. These were obtained by calculating the number of trajectories crossing each grid square and the contribution that each square made per trajectory to the wet and dry deposition in southern Norway. From this, and by grouping the squares into countries or regions, the figures shown in the tables could be deduced. It can be seen from table 3 that there is a considerable annual variation in the total deposition, the source of which can be ascribed to the variations in annual rainfall which occur. Rainfall at Lista in southern Norway, for instance, was about 15 % below the average for the previous 20 years in 1973 and about 10 % above in 1974.

TABLE 5. ESTIMATED PERCENTAGE CONTRIBUTIONS TO TOTAL SULPHUR DEPOSITION IN NORWAY FOR 1974 GIVEN BY SMITH & JOHNSON (1975) AND O.E.C.D. (1977)

	percentage contribution	
country	Smith & Johnson	O.E.C.D.
British Isles	37.2	28.0
Norway	17.8	12.0
G.D.R.	7.2	4.0
F.R.G.	10.4	4.0
Sweden	5.1	3.6
Holland	3.3	1.6
Poland	3.4	2.0
Denmark	2.3	3.2
France	3.4	3.6
Belgium	3.0	1.6
Czechoslovakia	1.6	1.2
U.S.S.R.	2.3	—
Italy	1.1	0
Finland	0.9	0.4
others	1.0	1.6
undecided	—	40.0

Over longer periods it might be expected that the annual rainfall amounts will vary between about 50 % and 150 % of the long-term average at any particular site. Any trend in annual rainfall over a period of years will of course lead to a corresponding trend in sulphur deposition for the years in question. Table 3 shows that rainfall occurring with the prevailing southwesterly winds cause the British Isles to be responsible for the biggest single contribution to wet deposition although again there are considerable yearly variations.

A different picture emerges with regard to dry deposition. The Norwegians themselves provide the largest contribution, showing the importance of local sources, and the frequency of southwesterly winds are again shown up in the British Isles figures. Table 4 shows the mean percentage of the total deposition per year contributed by each country. With the use of this model, then, Britain is responsible for about 30 % of the total deposition in Norway during the three years studied.

Table 5 compares the results of the model with results given in O.E.C.D. (1977) for a single year 1974. The latter are claimed to be accurate to within about ± 50 % and are for the whole of Norway while the former refer only to the southern part of the country. The same pattern can be seen in both sets of results although in the O.E.C.D. column, a disturbingly large amount (40 %) of the deposition was from 'undecided' origins. This arises in part from the way N.I.L.U.

estimated rainfall from a comparison of model-calculated concentrations of total sulphur with observed concentrations of sulphate in precipitation which implies a non-zero wet deposition even when air sulphur concentrations fall to zero. The causes of this apparent anomaly are not fully understood but one possibility is that local sources are making an inadequate contribution in the N.I.L.U. model. In Norway much of the rain is associated with westerly flow from over the Atlantic without any upwind input of sulphur, and the N.I.L.U. technique gives an apparent sizeable wet deposition without any clear origins. The Meteorological Office model used actual rainfall observations to assess wet deposition and so avoided this problem of an 'undecided' source. When considering amounts of wet deposition it is interesting to note that the correlation between rainfall and sulphate wet deposition is not particularly high on a day-by-day basis. Indeed in some areas it has been found to be as low as 0.2 although more normally it is 0.6–0.7. This means that it is not advisable to look *only* at rainfall distribution in order to draw conclusions about the wet deposition pattern. The reason for this is in the position of each area relative to the main emission areas. Sites just to the west of major source areas in Europe, for instance, will only receive small concentrations of sulphate in most of their heavier rainfall, which will tend to occur when the winds are westerly in direction. However, easterly winds, although less common and producing relatively small rainfalls, will tend to be rather heavily polluted. This means that at such a site wet deposition tends to be uniform in magnitude and has a quite different magnitude–frequency distribution from the rainfall. The importance of this becomes apparent in the next section.

Episodes

In some parts of Europe a significant proportion of the total annual wet deposition occurs on only a few days of the year. We can describe this feature quantatively by defining 'episode-days' at a given point as those days with the highest wet depositions which, when summed, account for 30 % of the total annual wet deposition and then define the 'episodicity' at that place as the ratio of the number of episode-days to the total number of rainfall days (figure 5 gives an example of these definitions. The episodicity at Cottered in the east Midlands is 5.3 % from 1974 data). Using these definitions it can be shown that, for 1974 data, many parts of central and southern Norway, as well as parts of south Switzerland and Scotland, have episodicities of less than 5 %. These areas are distant from the main industrial regions and the air is normally unpolluted. Just occasionally, however, air arriving there will have crossed a source area some time before, and if this air is introduced into frontal rain or thunderstorms, high depositions can occur. In these 'highly episodic' areas the number of episode-days is typically less than 10 and can be as low as 4, i.e. 30 % of the total annual wet deposition falls on 4 days.

The importance of the effect of episodes on the environment is not entirely obvious at present, although it is possible that occasions of significant wet deposition of unusually low pH may result in marked fish kill in streams and small lakes. The relevance of episodes with regard to any possible control of emissions is however, clearer. If the mechanisms leading to episodes could be sufficiently understood to be included in models, it would be possible for budgeting during episode-days to be carried out. However, the very nature of episodes precludes the possibility of them being adequately dealt with by statistical models of the type described earlier. These isolated events will have to be treated individually and may lead to significantly different results. It has been observed that most episodes are associated either with slow-moving frontal systems where heavy and prolonged rainfall can be produced by convergence of moist air at low levels, often enhanced by mountains, or with a slow-moving anticyclone centred over a major source area. If, in this latter case, air from near the centre of the anticyclone, which is by then heavily polluted, becomes drawn into a frontal zone, only relatively small amounts of rain can lead to high depositions.

Figure 6 shows back-trajectories for the 4 episode-days in 1974 at a site in southern Norway. It can be seen that all trajectories pass over major emission areas to the south or east. One of these

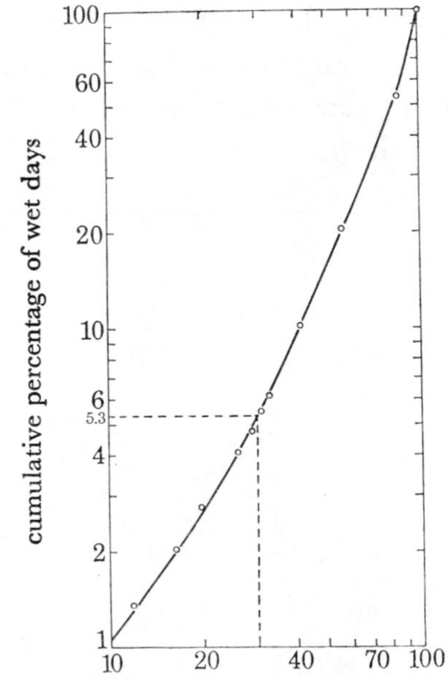

FIGURE 5. The cumulative percentage of deposition plotted against cumulative percentage of wet days, plotted on logarithmic scales, for Cottered in the east Midlands.

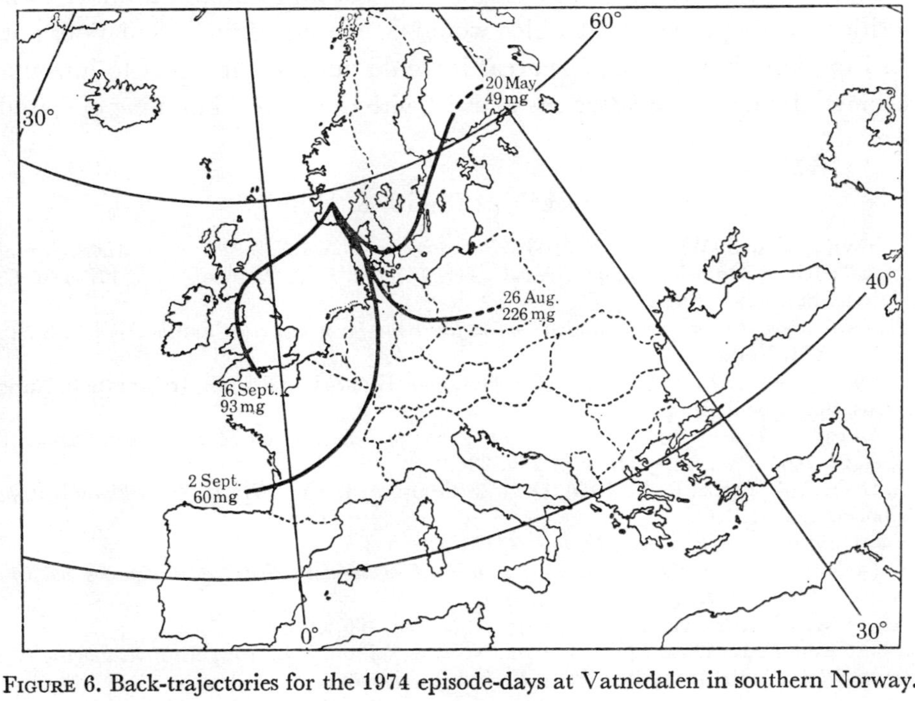

FIGURE 6. Back-trajectories for the 1974 episode-days at Vatnedalen in southern Norway.

trajectories, together with part of the emission grid and measured values of surface geostrophic wind, is shown in detail in figure 7. Some 48 h before the destination time, the trajectory was over a high emission area in the German Democratic Republic and Poland and close to a slowly moving anticylcone. As can be seen from the diagram, the surface geostrophic winds were only about 3–4 m s^{-1} allowing high concentrations of sulphur to accumulate. The polluted air was then drawn towards southern Scandinavia where a frontal system was approaching from the west. The increasing wind along the trajectory meant that only a relatively small amount of dry deposition took place, this loss to the ground being inversely proportional to the wind speed. Finally the air, which still contained high concentrations of sulphur, was drawn into the frontal zone over the mountains where large amounts of rain occurred with associated thunderstorms and produced by far the largest wet deposition of the year there.

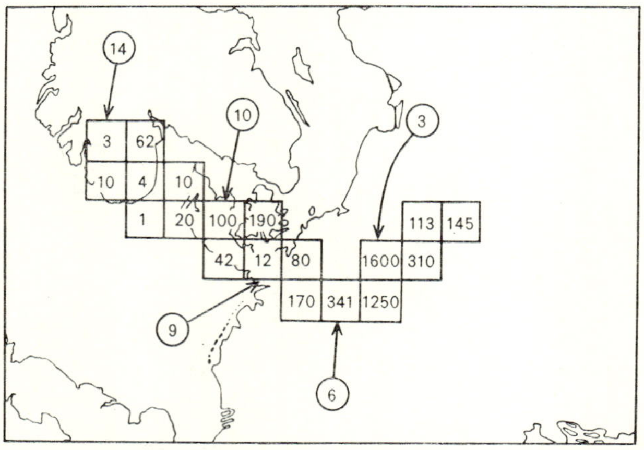

FIGURE 7. Back-trajectory for 0h 00 G.M.T., 27 August 1974, from Vatnedalen with relevant part of emission grid (figure 3). Number in circles refer to surface geostrophic winds (metres per second) along the route.

In fact, the rates of horizontal convergence which can be found near to thunderstorms can lead to wet depositions of about twice that which would be expected from calculations not including this feature. Especially in mountainous areas it would appear important to introduce convergence effects into any modelling attempts to assist with the correct handling of episodes.

REFERENCES (Smith & Hunt)

Allen, P. W., Jessup, E. A. & White, R. E. 1967 *Proceedings of U.S.A.E.C. Symposium Meteorological Information Meeting*, Chalk River, Ontario, Canada. A.E.C. L.-2787 (ed. C. A. Mawson), pp. 176–190. Chalk River: Atomic Energy of Canada Ltd.

Braekke, F. H. (ed.) 1976 *Impact of acid precipitation on forest and freshwater ecosystems in Norway*. S.N.S.F. project, NISK, 1432-NLH, Norway.

Briggs, G. A. 1969 *Plume rise*. TID-25075, Clearing House for Fed. Sci. and Tech. Information, National Bureau of Standards, Springfield, Virginia.

Browning, K. A. 1977 *The short period weather forecasting pilot project*. Internal Meteorological Office Paper.

Chamberlain, A. C. 1966 *Proc. R. Soc. Lond.* A **290**, 236.

Clarke, R. H., Dyer, A. J., Brook, R. R., Reid, D. G. & Troup, A. J. 1971 *The Wangara Experiment: boundary layer data*. Melbourne: C.S.I.R.O.

Hanna, S. R. & Gifford, F. 1973 *Atmos. Environ.* **7**, 131–136.

Heffter, J. L., Taylor, A. D. & Ferber, G. J. 1975 *A regional–continental scale transport, diffusion and deposition model.* N.O.A.A. tech. Memo E.R.L. ARL–50, 28 pages.

Hill, F. F., Whyte, K. W. & Browning, K. A. 1977 *Met. Mag.* **106**, 69–89.

Husar, R. B., Patterson, D. E., Husar, J. D., Gillani, N. V. & Wilson, J. E. 1977 In *Proceedings of the International Symposium on Sulphur in the Atmosphere. Atmos. Environ.* **12**, 549–568.

I.S.S.A. Husar, R. B. (ed.) 1977 *Proceedings of the International Symposium on Sulphur in the Atmosphere*, Dubrovnik. *Atmos. Environ.* **12** (1-3).
Izumi, Y. & Caughey, J. S. 1976 *Minnesota 1973 atmospheric boundary layer experiment data report.* Air Force Cambridge Research Laboratories Report Number AFCRL-TR-76-0038.
Moore, D. J. 1974 *Atmos. Environ.* **8**, 441–458.
O.E.C.D. 1977 *The O.E.C.D. programme on long range transport of air pollution.* Paris: O.E.C.D.
Pack, D. H., Ferber, G. J., Heffter, J. L., Telegadas, K., Angell, J. K., Hoecker, W. H. & Machta, L. 1977 In *Proceedings of the International Symposium on Sulphur in the Atmosphere. Atmos. Environ.* **12**, 425–444.
Scriven, R. A. & Fisher, B. E. A. 1975 *Atmos. Environ.* **9**, 49–68.
Smith, F. B. 1975 *Sci. Prog. Oxf.* **62**, 127–151.
Smith, F. B. 1977 *Application of data from field programmes to estimation of K-profiles and vertical dispersion.* Internal Meteorological Office paper.
Smith, F. B. & Hunt, R. D. 1977 In *Proceedings of the International Symposium on Sulphur in the Atmosphere. Atmos. Environ.* **12**, 461–478.
Smith, F. B. & Jeffrey, G. H. 1975 *Atmos. Environ.* **9**, 643–659.
Smith, F. B. & Johnson, D. A. 1975 *Sulphur transport in Western Europe during 1972–1974 from trajectory analysis and experimental measurements.* Internal Meteorological Office paper.
Sykes, R. I. & Hatton, L. 1976 *Atmos. Environ.* **10**, 925–934.
Weatherley, M.-L. P. M. 1977 *Fuel consumption, and smoke and sulphur dioxide emissions, in the United Kingdom, up to 1976.* Warren Spring Laboratory, Report LR 258 (AP).

Discussion

M. H. Unsworth (*University of Nottingham School of Agriculture, Sutton Bonington, Loughborough, Leics. LE12 5RD, U.K.*). Although there are a number of published measurements of deposition velocities for sulphur dioxide to natural surfaces, there have been few studies in winter when surfaces are often wet with dew and rain. Observations by Fowler (Fowler & Unsworth 1974; Fowler 1976) showed that when a wheat crop was wet with dew it had a much lower resistance to SO_2 uptake than when the crop was dry; a few measurements during rainfall led to similar conclusions (Fowler 1976). These observations suggest that deposition velocities for SO_2 on wet vegetation may be larger than the generally accepted values for dry surfaces. Would Dr Smith like to comment on the likely consequences of such a result on his calculations of long-range transport?

References

Fowler, D. & Unsworth, M. H. 1974 *Nature, Lond.* **249**, 389–390.
Fowler, D. 1976 Uptake of sulphur dioxide by crops and soil. Ph.D. thesis, University of Nottingham.

F. B. Smith. The consequences of a higher deposition velocity when the underlying vegetation is wet are not readily assessed. Two rather inadequate statements may, however, be made. Suppose the air behind a rain belt advects without actually catching it up and the sulphur dioxide it carries is affected by an enhanced deposition velocity along the whole of its track. As Professor R. S. Scorer once pointed out, the increased losses and subsequent lowering of air concentration tend to lead to a virtual independence of deposition rate on deposition velocity at ranges of about 1000 km from source. This distance corresponds well with the distance of the main European source regions from southern Norway. If, however, the air does gets advected into the rain belt, less sulphur will be there to be wet-deposited out, although this reduction will to some extent be minimized by the somewhat rapid oxidation of the sulphur dioxide to sulphate in the rather humid conditions along the tract and a consequent reduction in losses by dry deposition. Overall, then, some reduction in wet-deposition must be expected if Dr Unsworth's suppositions are correct.

R. A. Scriven (*Central Electricity Research Laboratories, Kelvin Avenue, Leatherhead, Surrey, U.K.*). As Dr Smith has described, it is now estimated that about 25 % of the sulphur deposited annually in Norway comes from the U.K. Although this is the largest contribution to this area from individual European countries, some further amplification of Dr Smith's figures is required to put the U.K. position properly into perspective.

In the O.E.C.D. study it was found that about 40 % of the deposition in Norway could not be allocated to any particular country because of the uncertainties associated with specifying wind trajectories accurately and because of the significant amount of sulphate in rain arriving on wind from the Atlantic sector where there were no industrial sources. In fact the average concentration of sulphate in rain from this sector was found to be one half of that from W–SW sector which contains the U.K. as the only major source. The corresponding sulphate in rain concentrations from the southerly and easterly sectors where the rain arrives in Norway on winds from the main continental source regions are three to four times that from the U.K. sector. The amount of rain coming from these sectors is about one-sixth of the total whereas the amount from the U.K. sector is about one-half. Hence, while the depositions from these various sectors are comparable the continental contribution consists of fewer occasion of higher concentrations as compared with the more frequent low concentration contribution from the U.K. sector. This distinction is likely to be important when discussing biological and geochemical effects.

The fact that a significant amount of sulphate arrives in Norway from winds in the Atlantic sector suggests that sulphur emitted from natural sources together with man-made emissions in North America (A. Nyberg 1977 *Q. Jl R. met. Soc.* **103**, 607–613) is travelling intercontinental distances as sulphate aerosol, the removal rate of which dry processes is known to be small.

P. F. Chester (*Central Electricity Research Laboratories, Kelvin Avenue, Leatherhead, Surrey, U.K.*). In introducing his paper, Dr Smith noted the assumption behind the O.E.C.D. survey that a major part of the acidity in Scandinavian rivers and lakes has its origins in rain borne sulphur compounds. In view of the extensive and detailed understanding of the transport and transformation of atmospheric sulphur exemplified by the last two papers, it is timely to contrast this body of knowledge with that on which the motivating assumption is based. Rosenqvist, in a recent (1976) report for the Norwegian Council of Scientific Research, discussed a number of biochemical and geochemical reactions having an important influence on Scandinavian inland waters. Until recently, the available chemical analysis have not included factors which would demonstrate the presence of absence of such mechanisms. However, in the spring of 1977, members of the Central Electricity Research Laboratories, working with scientists of the Norwegian S.N.S.F. Project, on the Tovdal River in Southern Norway, measured weak acid concentrations, which were presumably derived from interactions with the soil, up to 20 times those of the strong acids present. We estimate that the pH would have risen by only 0.2–0.3 if no strong acids had been present at all. It would appear that further work is necessary to elucidate the reaction paths leading to low pH in Scandinavian waters so as to identify more completely the responsible environmental factors. The results referred to will be reported in more detail elsewhere.

Reference

Rosenqvist, I. T. 1976 *Contribution to an analysis of the buffer effect of geological materials in respect of strong acids in rain water with examples from Numedal.* Report submitted to N.A.V.S. (Norwegian Council for Scientific Research).

Phil. Trans. R. Soc. Lond. A. **290**, 543–550 (1979)
Printed in Great Britain

Photochemical oxidation of atmospheric sulphur dioxide

By R. A. Cox
Environmental and Medical Sciences Division, A.E.R.E., Harwell, Oxfordshire OX11 0RA, U.K.

Oxidation of atmospheric sulphur dioxide can occur by homogeneous photochemically initiated gas-phase reactions as well as by heterogeneous reactions in cloud and fog droplets. Gas phase oxidation can result from reactions of excited SO_2 molecules formed by absorption of solar u.v. radiation by ground state SO_2, from reactions of SO_2 with photochemically generated OH and RO_2 free radicals, and from its reaction with transient species produced in thermal ozone–alkene reactions. Evaluation of the available mechanistic and rate data reveals that, of these three processes, oxidation by free radicals, particularly OH, is likely to be the most important in the atmosphere. Oxidation rates of up to 4 % h^{-1} are predicted for a hydrocarbon–NO_x polluted atmosphere under western European summertime conditions. This can lead to the formation of elevated concentrations of sulphuric acid and sulphate aerosol in polluted air. In the natural background troposphere oxidation rates are much less, *ca.* 0.3 % h^{-1} averaged over 24 h, but probably still significant as a source of atmospheric sulphates.

Introduction

Sulphur dioxide is one of the most widely recognized atmospheric pollutants, world-wide emissions amounting to approximately 150×10^6 t, 90 % of which originates in the Northern Hemisphere (Robinson & Robbins 1970). These man-made sulphur emissions are superimposed on emissions of natural gaseous sulphur compounds, mainly H_2S and organic sulphur compounds, but it is generally acknowledged that in populated and industrial regions, e.g. NW Europe, the man-made sources dominate. A significant fraction of all gaseous sulphur compounds are oxidized within the atmosphere to form sulphate, which can be detected in rain and as a component of the atmospheric aerosol in the form of sulphuric acid, acid and neutral ammonium sulphates, and other sulphates. The sulphates in atmospheric precipitation elements have environmental significance in their health effects (Higgins & Ferris 1973), their effects on visibility (Eggleton 1969) and in the ecological consequences of acid rainfall (U.N. 1971).

Over the years a number of mechanisms have been proposed to explain the oxidation of atmospheric SO_2 to H_2SO_4 and sulphates, and these may be broadly classified into two types, namely:

(*a*) homogeneous gas-phase oxidation, usually photochemically initiated;

(*b*) heterogeneous oxidation either in aqueous cloud or fog droplets or on aerosol particles.

The evidence seems to point to the occurrence in the atmosphere of both types of process, the contribution of each to the overall conversion of SO_2 to SO_4^{2-} depending on the prevailing ambient conditions. In the present discussion a brief survey of the various homogeneous, photochemically initiated mechanisms which have been proposed for SO_2 oxidation will be given. The quantitative conversion rates resulting from the more important processes will be discussed in terms of sulphuric acid formation in polluted air and the overall life-cycle of atmospheric SO_2.

Homogeneous gas-phase oxidation of SO_2

Three distinct mechanisms have been discussed for the homogeneous gas-phase oxidation of SO_2. They are:

(*a*) direct photooxidation involving the reactions of excited SO_2 molecules produced by the absorption of solar u.v. radiation in the first and second absorption bands of SO_2;

(*b*) oxidation of SO_2 by atoms and free radicals which are generated photochemically;

(*c*) oxidation of SO_2 by reactive intermediates in ozone–olefin reactions.

A summary of the available information on the mechanism and rates for each of these three processes is given below.

(*a*) *Direct photooxidation of* SO_2

The photodissociation of SO_2 by the reaction

$$SO_2 + h\nu = SO + O(^3P); \quad \Delta H^\circ_{298} = 547.6 \text{ kJ mol}^{-1}$$

requires light of wavelength $\lambda \leqslant 210$ nm. Therefore radiation in the solar u.v. at ground level ($\lambda \geqslant 290$ nm) is not sufficiently energetic to dissociate SO_2 and absorption in this region leads to the formation of electronically excited states of the SO_2 molecule:

$$SO_2 + h\nu \text{ (400–340 nm)} \longrightarrow SO_2(^3B_1), \text{ excited triplet } SO_2;$$

$$SO_2 + h\nu \text{ (340–260 nm)} \longrightarrow SO_2(^1B_1), \text{ and probably other singlet states.}$$

The photochemistry of SO_2 is therefore determined by the chemical behaviour of these excited states, the elucidation of which has proved to be a problem of some magnitude. Much of the available information on the excited states is due to the work of J. G. Calvert and coworkers, who have used fluorescence and phosphorescence emission to study the behaviour of the $^1B_1$ and $^3B_1$ states respectively (see, for example, Sidebottom *et al.* 1973) as well as chemical studies (Chung, Calvert & Bottenheim, 1975).

Absorption in the 400–340 nm 'forbidden' band is very weak and most excitation is initially into the singlet states. The singlet state is relatively unreactive chemically but is rapidly quenched by other gases including the major atmospheric components (N_2, O_2, H_2O, CO_2), e.g.

$$SO_2(^1B_1) + M \longrightarrow \underset{\text{ground state}}{SO_2(\tilde{X}\,^1A_1)} + M$$

$$\longrightarrow SO_2(^3B_1) + M.$$

A fraction of these quenching collisions lead to intersystem crossing to form the triplet state, which appears to be the dominant chemically active species in SO_2 photolysis systems. It can, for example, undergo chemical reactions with NO, SO_2 (to form SO_3), CO and hydrocarbons (to form sulphinic acids). The major atmospheric gases also quench SO_2 ($^3B_1$) to the ground state with moderate efficiency, however, and consequently the reactions of triplet SO_2 molecules with other trace constituents in the atmosphere are of negligible importance except under very special conditions. The only reactions of the $^3B_1$ state which can be significant in the normal atmosphere are possible chemical quenching reactions with O_2, e.g.

$$SO_3(^3B_1) + O_2 = SO_3 + O(^3P_1); \quad \Delta H = -159 \text{ kJ mol}^{-1}.$$

Although it is energetically favourable, this process is 'forbidden' by spin conservation rules, and might therefore be expected to be slow. Studies of the quantum yield for the photooxidation of SO_2 in the presence of excess O_2 (Cox 1972, 1973; Friend, Leifer & Trichon 1973) show this to be the case. Cox (1972) found for photolysis of SO_2 at high dilution in air,

$\Phi_{SO_3} \leqslant 3\times10^{-4}$, which is consistent with about 1% of the triplet SO_2+O_2 quenching encounters leading to chemical reaction to form SO_3.

The total rate of production of $SO_2(^3B_1)$ molecules in the atmosphere can be computed from the rate of absorption of sunlight by SO_2 in both the 'forbidden' and 'allowed' bands and the fraction of singlet SO_2 molecules which undergo intersystem crossing, approximately 10% for atmospheric conditions (Sidebottom *et al.* 1972). The relative quenching pathways for $SO_2(^3B_1)$ can also be computed from the available data and this allows an estimate of the total rate of reaction plus quenching for the $SO_2(^3B_1)+O_2$ interaction. For a solar zenith angle of 40° and 50% relative humidity the result is $2.6\pm0.8\times10^{-2}\ h^{-1}$. If only 1% of this reaction flux leads to SO_3 formation, the effective SO_2 photooxidation rate is $0.026\%\ h^{-1}$. As will be shown below, this is much slower than the oxidation rate resulting from the other homogeneous processes considered.

(*b*) *Oxidation of* SO_2 *by photochemically generated free radicals*

It can now be considered well established that the atmospheric oxidation of hydrocarbons and related substances proceeds by a photochemically initiated free radical chain process. The process occurs throughout the natural troposphere during daytime and, in polluted air, when the concentrations of trace components are large, it gives rise to the phenomenon of photochemical smog. The chain process is exemplified by the reactions involved in the oxidation of atmospheric methane to formaldehyde:

$$CH_4+OH \longrightarrow H_2O+CH_3,$$

$$CH_3+O_2 \xrightarrow{(M)} CH_3O_2,$$

$$CH_3O_2+NO \longrightarrow CH_3O+NO_2,$$

$$CH_3O+O_2 \longrightarrow HCHO+HO_2,$$

$$HO_2+NO \longrightarrow NO_2+OH.$$

The chain-carrying radicals are OH, HO_2 and their organic analogues RO and RO_2. In general, R can be alkyl, acyl or any other carbon containing fragment. The nitrogen oxides are involved through their ability to convert RO_2 (or HO_2) to RO (or OH) in the chain sequence and also through their rôle as a source of ozone, via NO_2 photolysis:

$$NO_2+h\nu(\lambda \leqslant 398\ nm) \longrightarrow NO+O(^3P_1),$$

$$O(^3P_1)+O_2 \xrightarrow{(M)} O_3.$$

Ozone in turn is important as a source of OH radicals through its photodissociation to yield electronically excited oxygen atoms which react with water vapour:

$$O_3+h\nu\ (\lambda \leqslant 310\ nm) = O(^1D_2)+O_2,$$

$$O(^1D_2)+H_2O = 2OH.$$

Since the free radical reactions are relatively rapid a 'steady-state' concentration of atoms and radicals is achieved in the sunlight irradiated atmosphere. Any trace gas, e.g. SO_2, which can react with the free radicals, will be removed at a rate given by

$$-d[SO_2]/dt = k_r[SO_2]\,[X],$$

where k_r is the effective bimolecular rate constant for reaction between radical X and SO_2, at the appropriate temperature and pressure. Provided the fraction of radicals reacting with SO_2 is small, the radical 'steady state' will not be affected by this process. If [X] can be evaluated by computer modelling of the major atmospheric reactions or, preferably, by direct measurements, the removal rate of SO_2 can be determined from a knowledge of k_r. Recent critical appraisals of the available literature (Eggleton & Cox 1978; Calvert & McQuigg 1975; Calvert *et al.* 1978) show that, of the possible reactions of SO_2 with radical type species, reaction with OH is probably dominant, with significant contributions due to reactions with HO_2 and CH_3O_2. Acyl peroxy and higher alkylperoxy radicals do not appear to react rapidly with SO_2 at all. We will consider here, therefore, only the relevant reactions with the reactive species:

$$OH + SO_2 \xrightarrow{(M)} HOSO_2; \quad (1)$$

$$HO_2 + SO_2 \longrightarrow OH + SO_3; \quad (2)$$

$$CH_3O_2 + SO_2 \longrightarrow CH_3O + SO_3. \quad (3)$$

The peroxy radical reactions (2) and (3) are relatively straightforward O-atom transfer reactions yielding SO_3. In the atmosphere this product will hydrolyse extremely rapidly to form sulphuric acid aerosol (Goodeve, Eastman & Dooley 1934; Castleman *et al.* 1975). The rate constant k_2 has been measured (Payne, Stief & Davis 1973) by using a competitive kinetic technique and is reasonably well defined; $k_2 = 9 \times 10^{-16}$ cm^3 molecule^{-1} s^{-1} at 298 K, relative to the current consensus value for the reaction $HO_2 + HO_2 \rightarrow H_2O_2 + O_2$, $k = 3.5 \times 10^{-12}$ cm^3 molecule^{-1} s^{-1}. Reaction (3) has been investigated by observation of the decay of CH_3O_2 in the presence of SO_2 by the technique of flash photolysis with kinetic spectroscopy. Unfortunately, aerosol formation from the reaction products made accurate optical measurements difficult but a preliminary value of $k_3 \approx 5 \times 10^{-15}$ cm^3 molecule^{-1} s^{-1} has been reported (Whitbeck, Bottenheim, Levine & Calvert 1976).

The reaction of OH radicals with SO_2 is an addition reaction and knowledge of the fate of the $HOSO_2$ free radical formed initially is important, as well as the overall rate constant and its pressure dependence. Some of the first investigations of this reaction at room temperature and atmospheric pressure were made by using the photolysis of nitrous acid as a source of OH radicals (Cox 1975). This study provided an estimate of k_1 relative to the rate constant for the reaction of OH with CO, and also showed that in moist air the $HOSO_2$ radical reacts to form ultimately a sulphur-bearing aerosol, similar in character to model sulphuric acid aerosols. In the presence of relatively high NO_2 concentrations, the aerosol also contained an NO_x component which hydrolysed in solution to yield NO_3^- in equimolar proportion to H_2SO_4 (Cox & Derwent 1976). Another observation of significance in this system was that oxidation of NO to NO_2 occurred in the reactions after the addition of OH to SO_2. These results provide good evidence for the occurrence of the following exothermic reactions which have been postulated by others (e.g. Calvert & McQuigg 1975) on thermochemical grounds:

$$HOSO_2 + O_2 = HOSO_2O_2;$$

$$HOSO_2O_2 + NO = HOSO_2O + NO_2;$$

$$HOSO_2O + NO_2 = HOSO_2ONO_2 \text{ (nitrylsulphuric acid)}.$$

There is at present, however, no satisfactory mechanism for the conversion of the $HOSO_2O$ radical to SO_3 or H_2SO_4 under all the experimental conditions.

Because of the mechanistic uncertainties in the above system, k_1 could only be evaluated with limited accuracy; based on the 'high pressure' value of 3×10^{-13} cm^3 molecule^{-1} s^{-1} for the rate constant for the reaction of OH with CO, (Cox, Derwent & Holt 1976), $k_1 \approx 5 \times 10^{-13}$ with an uncertainty of $\pm 50\,\%$ (Cox & Derwent 1976). Castleman & Tang (1977), who also used a competitive technique involving the OH + CO reaction, have demonstrated the expected pressure dependence of the association reaction (1) over the range 20–1000 Torr.† At 760 Torr N_2, k_1 was 1.4×10^{-12} (again relative to the 'high pressure' OH + CO rate constant). Direct measurements of k_1 by the molecular resonance fluoresence techniques to follow OH radical decay in the presence of SO_2 (Davis & Klauber 1975; Atkinson, Perry & Pitts 1976), give values of k_1 near to 8×10^{-13} cm^3 molecule^{-1} s^{-1} for pressure of N_2 near 760 Torr. Despite the experimental difficulties in this type of measurement in the presence of SO_2, these must be considered the most reliable values at present. It should be emphasized, however, that k_1 cannot be considered as well defined as the rate constants for the reactions of OH with other simple molecules of atmospheric interest.

(*c*) *Oxidation of* SO_2 *by reactive intermediates in ozone–olefin reactions*

At the low concentrations of SO_2 and ozone present in the atmosphere, the exothermic reaction

$$SO_2 + O_3 = SO_3 + O_2; \quad \Delta H = -241.6 \text{ kJ mol}^{-1}$$

is, surprisingly, negligibly slow. However, in the presence of alkenes, which react relatively rapidly with ozone, appreciable oxidation of SO_2 occurs in the dark (Cox & Penkett 1971). The kinetics of SO_2 oxidation were well described by a mechanism involving reaction of SO_2 with a reactive intermediate, which was competitively removed by an undefined, pseudo first order process (Cox & Penkett 1972). It was suggested that the oxidizing species might be the biradical or 'zwitterionic' Criegee intermediate, RC^+HOO^-, or possibly the original 'molozonide' product of the addition of an ozone molecule to the olefinic double bond:

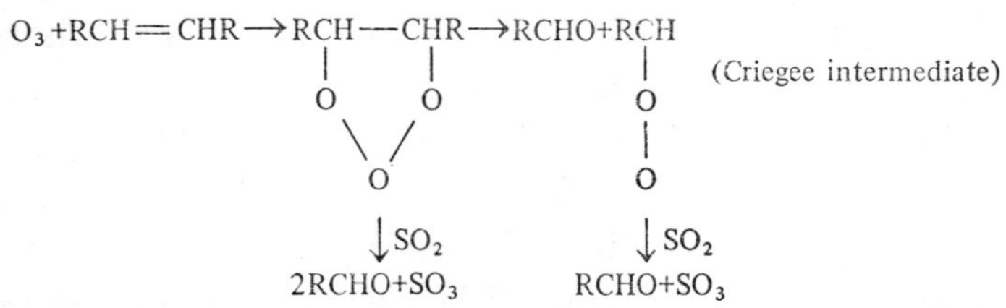

The observed increase in aldehyde products from the reaction in the presence of SO_2 provides support for the O atom transfer reactions suggested, which could occur in competition with fragmentation processes giving other products.

The mechanism of the gas phase reaction of ozone with alkenes is exceedingly complex and although recent work has provided more information on the kinetics of the reaction and on the products which are formed, a satisfactory mechanistic description of all of the experimental observations has not been achieved to date (see, for example, O'Niel & Blumstein 1973). The ozone–alkene–SO_2 system has also been investigated by other workers (McNelis 1974; Schulten & Shurath 1975). The results substantially confirm the original findings of Cox & Penkett but do not allow any firm conclusion regarding the nature of the oxidizing species. Irrespective of the nature of this species, the laboratory results giving the SO_2 oxidation rate as a function

† 1 Torr ≈ 133.3 Pa.

of reactant concentrations can be used as a basis for extrapolation to the atmosphere. These rates together with those estimated for the other homogeneous oxidation processes are summarized in the following section.

Table 1. Rates of homogeneous oxidation of SO_2 in the atmosphere

reaction	rate constant cm^3 molecule^{-1} s^{-1}	concentration of (active species) molecules cm^{-3}	100 × oxidation rate of SO_2 h^{-1}
(*a*) direct photooxidation			
$SO_2(^3B_1) + O_2 \rightarrow SO_3 + O(^3P)$	*ca.* 1.6×10^{-15}†	2.3 ($SO_2[^3B_1]$)‡	0.026
(*b*) free radical oxidation			
$OH + SO_2 \rightarrow HOSO_2$	8×10^{-13}	7×10^6§ (OH)	2.0
		(7×10^5)¶	(0.2)
$HO_2 + SO_2 \rightarrow OH + SO_3$	9×10^{-16}	2×10^9§ (HO_2)	0.6
		(1×10^8)¶	(0.03)
$CH_3O_2 + SO_2 \rightarrow CH_3O + SO_3$	5×10^{-15}	2×10^9§ (CH_3O_2)	1.7
		(1.5×10^7)¶	(0.03)
(*c*) ozone–alkene reaction			
$RCHOO + SO_2 \rightarrow RCHO + SO_3$	—	2.5×10^{12} [O_3]	0.07‖
		2.5×10^{11} [C_3H_6]	

† 0.4% of total quenching of $SO_2[^3B_1]$ at 50% relative humidity.
‡ Solar zenith angle = 40°; [SO_2] = 10 parts/10^9.
§ Typical maximum daytime OH and RO_2 concentrations computed for 50° N latitude, 1 July; see text.
‖ Extrapolated from laboratory measurements of O_3–propene–SO_2 system.
¶ Mean summertime concentrations in the background troposphere at 50° N latitude.

Rate of homogeneous oxidation of SO_2 in the atmosphere

Table 1 shows a summary of the estimated rates of oxidation of SO_2 in the atmosphere based on the experimental studies discussed above for the three types of homogeneous oxidation mechanism. It will be seen that the direct photooxidation is very much slower than the free radical processes and can therefore be neglected in determining overall sulphate formation rates. The concentrations of free radical species OH, HO_2 and CH_3O_2 used to calculate the respective SO_2 oxidation rates are derived from a photochemical model of an urban air-shed, 1 km in height, irradiated by sunlight of intensity corresponding to 12 noon on July 1 at 50° N latitude (R. G. Derwent & R. A. Cox, unpublished work). The concentration of CO, NO_x, SO_2 and C_1–C_5 hydrocarbons in this model were typical average values taken from pollutant measurements in London and should be fairly representative for most European cities. The SO_2 oxidation rates calculated on this basis indicate almost equal contributions from reaction with OH and peroxymethyl radicals, with a lesser contribution from HO_2. By taking propene as a representative alkene in urban air, the contribution of the ozone–alkene–SO_2 oxidation process is also small compared to the free radical processes. However, in the presence of more reactive alkenes, which would of course only have a short atmospheric lifetime, faster SO_2 oxidation would occur, e.g. 1.4% h^{-1} with 10 parts/10^9† as Butene 2 + 100 parts $O_3/10^9$. Thus this process is almost certainly significant in some circumstances, in particular since it can operate during night-time and wintertime when photochemical processes are either absent or occur at much reduced intensity.

† 1 part/10^9 = 2.45×10^{10} molecules cm^{-3} at 298 K, 760 Torr.

The total SO_2 oxidation rate during the midday hours in summer is estimated to be 4 % h^{-1}, on the basis of the results in table 1. From typical urban SO_2 concentrations at this season, 50 parts/10^9 (Department of Trade and Industry 1972), the corresponding rate of sulphuric acid formation is 8.4 $\mu g\, m^{-3}\, h^{-1}$. Thus photochemical oxidation is expected to generate significant amounts of sulphuric acid in urban air, with important consequences for health and visibility.

In order to determine a realistic lifetime of SO_2 in the atmosphere with respect to homogeneous oxidation, oxidation rates in unpolluted air are also required, since the local removal processes are not rapid enough to prevent transport of SO_2 into the main tropospheric air movements. Background tropospheric concentrations of free radicals can also be computed using photochemical models.

Derwent & Curtis (1977) have recently reported a two-dimensional tropospheric model from which diurnally and seasonally averaged OH, HO_2 and CH_3O_2 concentrations can be derived. The 24 h mean concentrations of these radicals obtained from the model and the corresponding oxidation rates of SO_2, for summertime conditions at 50° N, are shown in parentheses in table 1. The total oxidation rate of $0.26 \times 10^{-2}\ h^{-1}$ corresponds to an SO_2 lifetime with respect to homogeneous oxidation of 16 days. This is rather long compared with the estimated life-time of SO_2 in the mid-latitude troposphere (3–6 days), owing to the other removal processes, dry deposition and oxidation and removal by precipitation elements. The latter would be even more important in the winter months when photochemical reactions become very slow because of the much reduced insolation at midlatitudes. It is therefore concluded that photochemical oxidation plays a minor but significant rôle in the overall oxidation of SO_2 in the natural unpolluted troposphere at mid-latitudes. In polluted urban plumes, and particularly during the summer months, quite rapid homogeneous photochemically initiated oxidation of SO_2 is predicted to occur.

References (Cox)

Atkinson, R., Perry, R. A. & Pitts, J. N. Jr 1976 *J. chem. Phys.* **65**, 306.

Calvert, J. G. & McQuigg, R. D. 1975 *Int. J. chem. Kinet.* (Symp. Issue) **1**, 113.

Calvert, J. G., Fu Su, Bottenheim, J. W. & Strauss, O. P. 1978 *Atmos. Environ.* (Sulphur Symp. Issue). (In the press.)

Castleman, A. W. Jr, Davis, R. E., Munkelwitz, H. R., Tang, I. N. & Wood, W. P. 1975 *Int. J. chem. Kinet.* (Symp. Issue) **1**, 629.

Castleman, A. W. & Tang, I. N. 1977 *J. Photochem.* **6**, 349.

Chung, K., Calvert, J. G. & Bottenheim, J. W. 1975 *Int. J. chem. Kinet.* **7**, 161.

Cox, R. A. 1972 *J. Phys. Chem.* **76**, 814.

Cox, R. A. 1973 *J. Aerosol Sci.* **4**, 473.

Cox, R. A. 1975 *J. Photochem.* **3**, 291.

Cox, R. A. & Derwent, R. G. 1976 Paper presented at COST 61A Technical Symposium, Ispra, Italy.

Cox, R. A., Derwent, R. G. & Holt, P. M. 1976 *J. chem. Soc. Faraday Trans.* I **72**, 2031.

Cox, R. A. & Penkett, S. A. 1971 *Nature, Lond.* **230**, 321.

Cox, R. A. & Penkett, S. A. 1972 *J. chem. Soc. Faraday Trans.* I **68**, 1735.

Department of Trade and Industry 1972 *National Survey of Air Pollution*, 1961–1971. London: H.M.S.O.

Derwent, R. G. & Curtis, A. R. 1977 A.E.R.E. Report R8853. London: H.M.S.O.

Davis, D. D. & Klauber, G. 1975 *Int. J. chem. Kinet.* (Symp. Issue) **1**, 543.

Eggleton, A. E. J. 1969 *Atmos. Environ.* **3**, 355.

Eggleton, A. E. J. & Cox, R. A. 1978 *Atmos. Environ.* (Sulphur Symp. Issue) **12**, 227.

Friend, J. P., Leifer, R. & Trichon, M. 1973 *J. atmos. Sci.* **30**, 465.

Goodeve, C. F., Eastman, A. S. & Dooley, A. 1934 *Trans. Faraday Soc.* **30**, 1127.

Higgins, I. T. T. & Ferris, B. G. Jr 1973 *Proc. Conf. on Health Effects of Air Pollutants*, p. 227. Washington, D.C.: U.S.G.P.O. Prepared for U.S. Senate Committee on Public Works by N.A.S./N.R.C.

McNelis, D. N. 1974 *E.P.A. Environmental Monitoring Series report* 650/4-74-034. Research Triangle Park: U.S Environmental Protection Agency.
O'Niel, H. E. & Blumstein, C. 1973 *Int. J. chem. Kinet.* **5**, 397.
Payne, W. A., Stief, L. J. & Davis, D. D. 1973 *J. Am. chem. Soc.* **95**, 7614.
Robinson, E. & Robbins, R. C. 1970 *J. Air Pollut. Control Ass.* **20**, 303.
Schulten, H. R. & Shurath, U. 1975 *J. phys. Chem.* **79**, 51.
Sidebottom, H. W., Badcock, C. C., Jackson, G. E., Calvert, J. G., Reinhardt, G. W. & Damon, E. K. 1973 *Environ. Sci. Technol.* **6**, 72.
U.N. 1971 *Air pollution across national boundaries* (Sweden's case study for the U.N. Conference on the Human Environment). Stockholm: Royal Ministry of Foreign Affairs, Royal Ministry of Agriculture.
Whitbeck, M. R., Bottenheim, J. W., Levine, S. Z. & Calvert, J. G. 1976 *Abstracts of 12th informal conference on photochemistry*, N.B.S. Gaittersburg Maryland, pp. K1-1 to K1-5.

Discussion

M. F. R. Mulcahy (*Atmospheric Chemistry Section, C.S.I.R.O. Division of Process Technology, P.O. Box* 136, *North Ryde, N.S.W., Australia* 2113). Dr Cox quoted measurements of the rate of atmospheric oxidation of sulphur dioxide made under various conditions and has given his opinion as to what basically is likely to be the reaction mechanism in air containing other urban pollutants. I should like to mention some results of a study of the kinetics of oxidation in a 'remote' atmosphere which is being carried out by my colleagues Mr D. J. Williams and Mr D. B. Roberts. The investigation is conducted over Northern Australia where, near the town of Mt Isa (lat. 21° S, population 25000), the two adjacent chimneys of a metallurgical smelter constitute a strong point source of SO_2. Apart from a small coal-burning power station, there is no other industrial activity at Mt Isa. When the meteorology is favourable, a well-defined plume from the smelter flows NW over several hundred kilometres of almost uninhabited country. Thus the sulphur dioxide is in contact with air as little contaminated by other sources of pollution as perhaps is likely to be found for such an investigation anywhere.

So far, Mr Williams has studied the rate of disappearance of SO_2 from the plume by two methods. The first method makes use of a Cospec correlation spectrometer to determine the total flux of SO_2, and the second uses a photometric instrument to determine the concentration ratio $[SO_2]$:[total S] in the plume. Both instruments are carried (separately) in an aircraft, and measurements have been made successfully at various distances up to about 250 km from the source.

The total flux method has yielded an upper limit of 1 % h^{-1} for the total rate of loss of SO_2 from the plume; that is, by oxidation and ground absorption combined.

More precise results have been obtained by the ratio method. They can be summarized as follows. There is a strong diurnal variation in the rate of loss of SO_2: 8–10 % of the SO_2 disappears from the plume during one period of daylight (that is, 10 h) but no loss can be detected at night. The rate of deposition of SO_2 to the ground has been measured by special experiments made on location. After subtracting this from the total loss from the plume, Mr Williams estimates the rate of atmospheric oxidation of SO_2 under the prevailing conditions to be 0.6 % h^{-1} averaged over the hours of daylight or 0.25 % h^{-1} averaged over 24 h.

I believe this to be the first unequivocal experimental demonstration that the oxidation of SO_2 in the free atmosphere is indeed principally photochemical.

Phil. Trans. R. Soc. Lond. A. **290**, 551–576 (1979)
Printed in Great Britain

Photochemical and biological implications of the atmospheric reactions of amines and benzo(a)pyrene

By J. N. Pitts Jr
Statewide Air Pollution Research Center and Department of Chemistry, University of California, Riverside, California 92521, *U.S.A.*

The discovery of direct mutagenic activity, as determined by the Ames *Salmonella typhimurium* reversion assay, in the organic fractions of ambient aerosols collected throughout Southern California, led us to investigate the reactions of benzo(a)pyrene (BaP) deposited on glass fibre filters with ambient photochemical smog, as well as with O_3, NO_2, and peroxyacetyl nitrate (PAN) in simulated atmospheres. A variety of products are readily formed, including phenols, diphenols, dihydrodiols, etc. Directly mutagenic nitroderivatives are formed upon exposure of BaP (a carcinogen and pro-mutagen) and of perylene and pyrene (non-mutagens) to 1 part/10^6 of NO_2 and a trace of HNO_3 in air. If such reactions occur in urban atmospheres they may account in part for the 'excess' carcinogenicity (over BaP and certain other polycyclics) observed in organic particulates collected from smog and exhausts of spark ignition and diesel engines. However, such gas–solid interface processes may also have occurred on the filters commonly used in particulate sampling. Thus the possibility of 'filter artefacts' must be recognized.

Carcinogenic *N*-nitrosamines have been detected in air at or near several industrial plants. Therefore, reactions of their possible precursors, e.g. secondary and tertiary methyl and ethylamines with NO_x–HONO were studied both at levels of less than 1 part/10^6 in air in a 50 m^3 outdoor chamber and at levels around 4 parts/10^6 in a long path (720 m) outdoor infrared cell with the use of Fourier transform spectroscopic techniques. At concentrations less than 1 part/10^6 diethyl and triethylamine readily form photochemical smog (O_3, PAN, aerosols, etc.). Additionally, small but significant amounts of diethylnitrosamine $(C_2H_5)_2NNO$ are formed in the dark from diethylamine (DEA) but destroyed in sunlight. In contrast, $(C_2H_5)_2NNO$ is initially formed on irradiation of triethylamine (TEA)-NO_x mixtures, then photodecomposed. Other significant nitrogenous compounds formed in sunlight include dialkylnitramines (R_2NNO_2, a major product) and substituted amides; small amounts of acetamide are present in the particulate phase from DEA and TEA. Both $(CH_3)_2NNO_2$ and CH_3CONH_2 are carcinogens in animals, though less potent than nitrosamines; the activity of $(C_2H_5)_2NNO_2$ is not known.

Environmental implications of both systems are discussed.

1. Introduction

Historically, researchers and control officials have focused their attention on the major air pollutants: sulphur dioxide, carbon monoxide, nitrogen dioxide, photochemical oxidant (including ozone) and total suspended particulates. In the United States these are often referred to as 'criteria pollutants'; thus, during the last decade the U.S. Environmental Agency (E.P.A.) has set Federal Ambient Air Quality Standards based to a large extent on the Air Quality Criteria documents developed for each pollutant, and has implemented strategies for their control.

Recently, however, considerable interest has developed in better understanding the sources, transport, and associated chemical and physical transformations, and sinks of several types of 'non-criteria' pollutants. Certain of these species, though present in urban atmospheres at much lower concentrations than the criteria pollutants, are known to produce disproportionately severe health effects in experimental animals – and possibly in man. A classic example is the polycyclic aromatic hydrocarbon (PAH) benzo(a)pyrene (BaP). It is formed during the combustion of fossil fuels and was identified by its fluorescence spectrum in domestic soot in 1949 (Goulden & Tipler 1949). In 1952 it was identified and measured in ambient soot particles collected at ten stations throughout Great Britain (Waller 1952). Today it is recognized as being distributed globally.

A second class of 'non-criteria' pollutants of current concern are the *N*-nitrosamines. Recently, they have been identified in a variety of systems including pesticides, synthetic cutting and grinding fluids, and drinking water (Fine 1978). However, in contrast to the ubiquitous nature of BaP, they have been identified as air pollutants only in a limited number of locations, including those in or near industrial plants in East Germany (Bretschneider & Matz 1973) and the U.S. (Fine, Rounbehler, Belcher & Epstein 1976; Fine *et al.* 1977; Pellizzari 1977; U.S. E.P.A. 1977). These industrial plants employed dimethylamine (DMA) or dimethylnitrosamine (DMN) respectively in organic syntheses.

The carcinogenic activity of soot particles has been of concern for over two centuries (Pott 1775) and that of *N*-nitrosamines since it was first shown in 1956 (Magee & Barnes 1956). Today the possibility that environmental contaminants may be responsible for a significant fraction of cancer in man is being widely discussed (Epstein 1974; Heidelberger 1975; Searle 1976; de Serres 1976*a, b*). Thus, while to date no direct link has been established unequivocally between urban air pollution and lung cancer there is substantial indirect evidence that a relation may exist (National Academy of Sciences 1972). Quantitative studies of the atmospheric chemistry of PAH and nitrosamines and their precursors therefore seem desirable.

Pronounced societal interest in these classes of compounds is illustrated by the amendments to the U.S. Clean Air Act passed by Congress and signed into law by President Carter in August 1977. Section 104b requires the Administrator of the Environmental Protection Agency (E.P.A.) to review and critique all available information on 'nitric and nitrous acids, nitrites, nitrates, nitrosamines and other carcinogenic and potentially carcinogenic derivatives of oxides of nitrogen'. Similarly in Section 122(a), the Administrator is required to 'determine whether or not emissions of polycyclic organic matter (POM) into the ambient air will cause, or contribute to air pollution which may reasonably be anticipated to endanger public health'.

The highly specific nature of these Congressional mandates is historically somewhat unusual in federal air pollution legislation. Clearly, they reflect growing concern over increasing emissions of oxides of nitrogen (NO_x) and possible undesirable consequences of proposed major shifts to a far greater use of coal, shale oil, diesel powered motor vehicles, etc., to relieve in part the energy crisis in the U.S.

This charge of Congress to the E.P.A. poses a variety of difficult scientific problems. For example, to establish the nature and magnitude of a health effect, one needs reliable dose–response curves for the air pollutant(s) in question. Elucidation of the 'dose', of course, requires detailed knowledge of the physical (i.e. gas, liquid, solid, adsorbed on solid, particle size, etc.) and chemical nature of the pollutant (and co-pollutants) at the site of its impact on the biological target. In addition, the actual 'dose' administered to man or animals may not necessarily be chemically the same as that determined by using current sampling and analytical techniques, a point illustrated in §2.6, where filter artefacts are discussed.

Central to the issue, as the title of this Discussion implies, is the fact that during transport in the atmosphere in the presence of sunlight, oxygen, water and a spectrum of co-pollutants, primary organic pollutants undergo a variety of chemical and physical transformations. Consequently, the mixtures of gaseous and particulate pollutants which impact urban man are complex, often containing hundreds of species. Furthermore, some pollutants of interest often occur at concentrations in the range of 1 part/10^9 or less. Thus detailed analysis for chemical carcinogens in ambient particulates, e.g. by gas chromatography – mass spectrometry (g.c.–m.s.) is complicated at best.

Compounding the problem is the fact that there not only is a dearth of reliable information in the literature but also a significant amount of misinformation on atmospheric transformations of 'non-criteria' pollutants. For example, clouding the issue with respect to the PAH is the statement that, as air pollutants, 'They are chemically inert and thus are removed from the air only by rain or the slow sedimentation of the particulate' (Berry & Lehman 1971; Fishbein 1976, p. 232). Actually, although the evidence to date is limited (see below) it is clear that many PAH, including BaP, are highly reactive compounds. These should undergo a variety of atmospheric reactions, thermal and photochemical, with a number of co-pollutants, both molecular and free radical species. The problem is that little is known about these processes (for detailed reviews see National Academy of Sciences (1972) and Hoffmann & Wynder (1977)).

With respect to nitrosamines, a recent E.P.A. document (1976) states that in solution 'Unlike the secondary amines which can form nitrosamines – only a few tertiary amines form nitrosamines.' This is clearly not the case; tertiary amines react readily in solution with aqueous nitrous acid to form secondary *N*-nitrosamines (Guether 1864; Smith & Loeppky 1967; Walters, 1977). The real question is whether or not tertiary, as well as secondary, amines can react with NO_x–HONO in the gas phase to form *N*-nitrosamines.

In this paper are presented current results from our laboratory dealing first with reactions in simulated and real urban atmospheres of the carcinogen and mutagen BaP and of its non-mutagenic (in the Ames reversion assay) isomer perylene, and second with reactions of mixtures of simple aliphatic amines and NO_x in air containing traces of nitrous acid (HONO). We shall also discuss the mutagenic activity of ambient urban particulates and show how the Ames test, which employs the bacterium *Salmonella typhimurium* to test for mutagens, can provide microbiological clues useful in establishing the chemical identity of the species responsible.

Although at first glance the two systems, particulate PAH and gaseous nitrosamines, may seem quite unrelated, they both participate in chemical transformations in which hydroxyl (OH) and hydroperoxyl (HO_2) radicals and NO_x play major rôles – just as they do in tropospheric photochemical smog and in determining the stratospheric ozone balance, the two general topics of this Discussion. For detailed considerations, the full papers should be consulted.

2. Atmospheric chemistry of PAH

2.1. *Carcinogenicity of organic particulates*

Domestic soot was shown to be carcinogenic over 50 years ago (Passey 1922), and it has been known for over three decades that extracts of the organic fraction of particulates collected in ambient urban air in the U.S. are carcinogenic when administered subcutaneously to mice (Leiter, Shimkin & Shear 1942; Leiter & Shear 1943). Subsequently, this effect was seen in experimental animals administered extracts from ambient particulates collected in 1954 from

Los Angeles photochemical smog (Kotin, Falk, Mader & Thomas 1953) and in seven U.S. cities (Hueper *et al.* 1962). Similar results have been found with ambient samples collected in major urban centres throughout the world.

The carcinogenicity of the neutral fraction of organic particulates has been mainly attributed to the presence of PAH such as BaP and benz(a)anthracene. Recently, by using fluorescence and/or combined g.c.–m.s. techniques, these compounds and a large number of other PAH have been identified in ambient particulates (Gordon & Bryan 1973; Lao, Thomas, Oja & Dubois 1973), motor exhaust (Boyer & Laitinen 1975), diesel engine exhaust (Spindt 1974) and in soil near mountain highways (Blumer, Blumer & Reich 1977). Indeed, the latter group suggested a correlation between the higher cancer death rates and the PAH emissions due to automobile traffic in the Swiss mountain area where their study was conducted. This has, however, been contested (Litman 1978).

In addition to PAH, other chemical carcinogens have been identified in ambient particulates, including aza-arenes such as benzocarbazoles in the neutral fraction, and benzacridines and dibenzacridines in the basic fraction (Sawicki *et al.* 1965; Sawicki 1967; Cautreels & Van Cauwenberghe 1976; Dong, Locke & Hoffman 1977; National Academy of Sciences 1972).

However, it is notable that the observed carcinogenicity in animals or transformation in cell cultures is significantly greater than can be accounted for by the amounts of polycyclics present in the samples administered (Kotin *et al.* 1954; Hueper *et al.* 1962; Epstein *et al.* 1966; Rigdon & Neal 1971; Freeman *et al.* 1971; Mohr *et al.* 1976; Grimmer 1977). Thus, Gordon and coworkers reported that in airborne particles collected in the Los Angeles area, the benzene extract had 100–1000 times the cell transformation activity attributable to its BaP content. Furthermore, the methanol extract, while containing only about one-thirtieth of the BaP in the total sample, had activity comparable to the benzene extract (Gordon *et al.* 1973).

With respect to auto exhaust, Mohr *et al.* (1976) studied the effects of the condensate on the Syrian golden hamster lung. They concluded not only that the condensate displayed a carcinogenic effect (100 % rate of multiple pulmonary tumours) but also 'Considering the relatively low total dose of BaP contained in the condensate, this pronounced neoplastic response cannot be explained alone by the effects of this well known carcinogenic hydrocarbon' (Mohr *et al.* 1976).

2.2. *The Ames test for mutagenic activity*

Because of the chemical complexity of ambient particulates and motor exhaust, and because animal tests for suspected carcinogens are time consuming and expensive, experiments directed to establishing the identity of the compounds responsible for the 'excess' carcinogenicity have been in general relatively limited. Recently, however, a rapid and relatively inexpensive microbiological assay for mutagenic activity has been developed by Ames and coworkers (Ames, Durston, Yamasaki & Lee 1973; Ames, McCann & Yamasaki 1975; Ames & McCann 1976). It is a reverse mutation system employing histidine-requiring mutants of the bacterium *Salmonella typhimurium*. In tests of some 300 compounds, about 85–90 % of the known carcinogens were also found to be mutagens, while the number of false positives and negatives ranged from 10 to 15 % (McCann, Choi, Yamasaki & Ames 1975). This observation has been confirmed by some investigators (Sugimura *et al.* 1976) but disputed by others (Ashby & Styles 1978). In any case, the Ames test is now generally recognized as a useful, *though by no means exclusive*, screening test for chemical mutagens in complex environmental samples (Bartsch

1976; Bochkov, Sram, Kuleshov & Zhurkov 1976; Commoner 1976; Dean 1976; de Serres 1976*a*, *b*; Sobels 1977).

We have been using the Ames test to screen for mutagenic activity in particulate samples collected from real and simulated atmospheres, and as a means of obtaining microbiological clues to the nature of the compounds responsible for the 'excess' carcinogenicity. Of course, we caution that the observation of mutagenic activity in bacterial assays conducted on environmental samples does not necessarily imply carcinogenic activity in man.

TABLE 1. DESCRIPTION OF AMES'S MICROBIOLOGICAL ASSAY FOR MUTAGENIC ACTIVITY

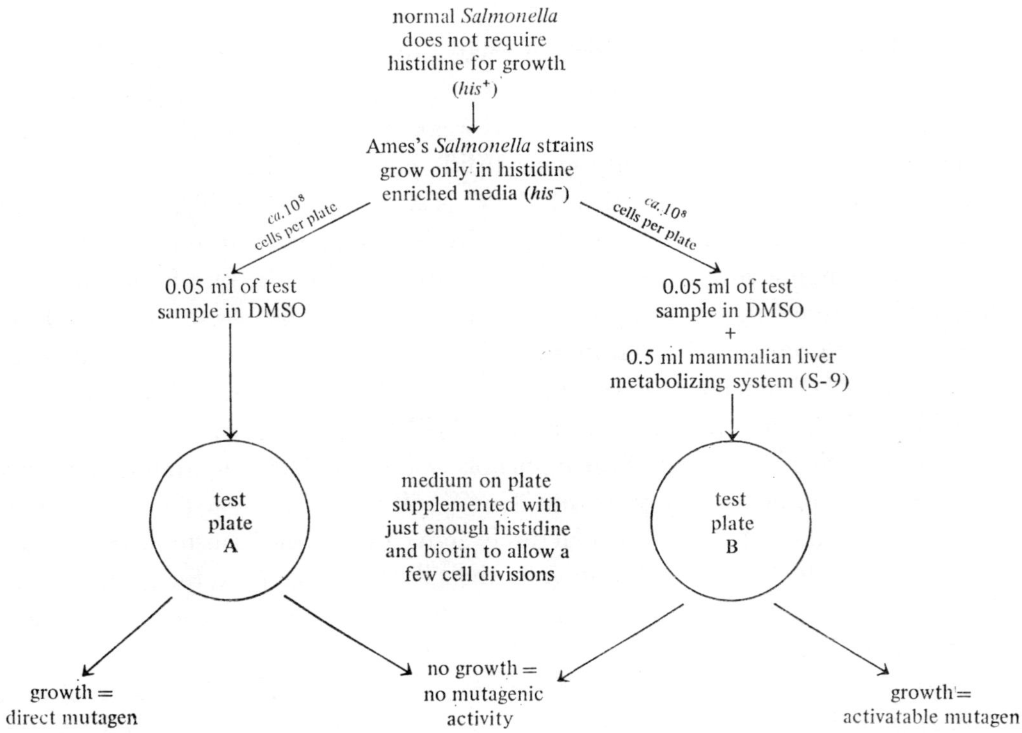

The experimental sequence used in the Ames test is outlined in table 1. In the test, one strain of *his*⁻ *Salmonella*, TA1535, is reverted to the normal *his*⁺ form with accompanying growth of colonies on the test plate by chemicals that are base pair substitution mutagens. Examples are *N*-methyl-*N*′-nitro-*N*-nitrosoguanidine and alkylating agents such as β-propiolactone. Other strains, TA1537, TA1538 and TA98, respond to chemicals which produce frameshift mutations such as aromatic amines (e.g. 2-aminofluorene and 2-acetylaminofluorene) and PAH such as BaP. Strain TA100 detects both base-pair substitution and frameshift mutagens.

Experimentally, one first adds portions of the test sample to a series of tubes, each containing agar and one of the test strains. These are then poured onto plates filled with minimal salts agar. If no significant increase of colonies above background (the background is caused by spontaneous *his*⁻ to *his*⁺ reversions) is observed for TA1535 but a pronounced growth in colonies is seen for TA1538 (or another frameshift sensitive strain, e.g. TA98), one concludes first that the sample is *directly* mutagenic and second that only chemicals causing frameshift mutations were present. Thus at least some of the compounds in the sample must be directly active frameshift mutagens.

This behaviour is in contrast to species such as BaP or 2-acetylaminofluorene which are *promutagens* (Lu *et al.* 1976); they are inactive *per se* and must be metabolized to other species before they can cause mutations in this bacterial assay. Microsomal activation is achieved by adding to the test sample a small amount of an enzyme system (S-9 solution, table 1, test plate B) derived from the livers of rats injected with Aroclor (a polychlorobiphenyl) or other such inducing agents.

Now, if we add S-9 and do *not* observe on plate B a significant increase in the number of TA1538 colonies above that seen on plate A we can conclude that *only* directly mutagenic frameshift chemicals are present. If, however, the number of revertants per microgram sample on B is significantly larger than on A, we can conclude that *both* direct and activatable (e.g. BaP) chemical mutagens are present.

Finally, the test can be highly sensitive to relatively small changes in chemical structure. Thus, for example, 2-acetylaminofluorene is a strong frameshift mutagen [108 revertants per nanomole (108 rev./nmol)] while the 4-isomer is weak (0.3 rev./nmol, McCann 1976).

The phenomenon of strain specificity toward different frameshift mutagens underlines the desirability of employing each strain (i.e. TA1535, TA1537, TA1538, TA98, TA100) at least once during initial screening tests for mutagenicity in complex environmental samples. For example, of the 12 monohydroxy-isomers of BaP, only five phenols were found to be directly acting frameshift mutagens, primarily 6-OH and 12-OH but also 1-, 3- and 7-OH-BaP to a smaller extent. The remaining seven isomers were non-mutagenic. However, the activity of the 6-OH and 12-OH isometers was reported to be weak when strain TA1538 was used but moderate with strain TA98 (Jerina *et al.* 1976*a*, *b*; Wislocki *et al.* 1976).

It is also worth noting that seven of these phenols (4-, 5-, 6-, 7-, 8-, 9-, and 10-hydroxy-BaP) were also tested for carcinogenicity on mouse skin (Kapitulnik *et al.* 1976); all proved negative. Furthermore, although 6-OH BaP was moderately active in Chinese hamster V79 cells, 12-OH-BaP was inactive (Wislocki *et al.* 1976). Such findings illustrate the need for caution if one uses conclusions based on results obtained from one particular kind of mutagenicity assay system to predict the behaviour of another type.

Finally, in view of its increasingly widespread use, the need for standardization of the Ames test, including plate incubation temperature (*precise* control at 37 °C is important (Winer *et al.* 1978)) as well as the number of cells per plate, plate volume, concentrations of rat liver S-9 homogenate, etc., is becoming increasingly urgent (Pitts *et al.* 1978*a*). Collaborative programmes to achieve this objective are currently under way.

2.3. *Mutagenic activity of urban organic particulates*

In 1975 the mutagenicity of organic fractions from particulates collected at several sites in the Los Angeles Basin was demonstrated using Ames's assay system (Pitts 1975). This phenomenon has now been reported in studies at Ohmura and Fukuoka, Japan (Tokiwa *et al.* 1976), Kobe, Japan (Teranishi, Hamada & Watanabe 1978), Buffalo, New York and Berkeley, California (Talcott & Wei 1977), New York City, New York (Daisey, Hawryluk, Kneip & Mukai 1978) and Chicago, Illinois (Commoner, Madyastha, Bronsdon & Vithayathil 1978).

Recently, as a means of obtaining more information about the chemical nature of the organic particulates and hence about atmospheric transformations of PAH, we expanded our programme and collected 23 samples of airborne particulates at 11 urban sites in California's South Coast Air Basin (Pitts *et al.* 1978*a*, *e*). Results of the Ames tests can be summarized as follows:

(i) All 23 samples exhibited frameshift-type mutagenic activity without requiring metabolic activation with strains TA1537, TA1538 and TA98.

(ii) All solvent and filter blanks were inactive.

(iii) Typically, assay of 0.1–2.0 mg of airborne organic particulates resulted in a 5- to 20-fold increase in the number of histidine revertants per plate above the background of spontaneous revertants in strains TA1537, TA1538 and TA98, and in up to a twofold increase in strains TA100.

(iv) No activity was observed in any of the assays with strain TA1535, which is reverted by base-pair substitution mutations.

(v) Microsomal activation by S-9 solution did not greatly increase the activity of most of the samples tested.

(vi) Assay of extracts of a size-resolved sample collected in downtown Los Angeles revealed that all mutagenic activity was associated with organic species present in particles of diameter 1.1 μm or less. This is consistent with the well documented distribution of organic particulate pollutants such as BaP with respect to particle size (Kertesz-Saringer, Meszaros & Varkonyi 1971; Natusch & Wallace 1974; Pierce & Katz 1975; Friedlander & Miguel 1978).

As noted earlier, the carcinogenic activity of organic fractions extracted from ambient particulates taken from the Los Angeles atmosphere is well known, so that our finding frameshift-type mutagenic activity in all urban samples is not surprising. What is interesting chemically is that metabolic activation was *not* required for any sample. Thus these urban particulates must contain mutagens other than the carcinogenic PAH, such as BaP, which require activation. This is consistent with the numerous observations of 'excess' carcinogenicity in animals or in cell transformation activity discussed above, and the low average concentrations of BaP in the Los Angeles Basin (Gordon & Bryan 1973).

2.4. *Atmospheric reactions of benzo(a)pyrene and perylene*

We recently proposed that the presence of these direct mutagens in ambient particulates may be due in part to the reactions of BaP and other PAH with species present in photochemical smog such as ozone, nitrogen dioxide, peroxyacetyl nitrate (PAN), singlet molecular oxygen ($O_2{}^1\Delta$), and the free radicals OH and HO_2 (Pitts *et al.* 1977*a*, 1978*c*). Some of the products formed in these reactions are analogous to the metabolites of BaP and other PAH in mammalian cells (Jerina *et al.* 1976*a*, *b*; Fahmy & Fahmy 1976), so that reactions of PAH in exhaust effluents and in the atmosphere could account for the presence of directly active mutagens in urban particulates.

Furthermore, we recently observed significant differences in the mutagenic activity of polar neutral fractions of ambient particulates sampled during daytime and nighttime (Pitts *et al.* 1978*b*). Since the concentrations of O_3, NO_2 and PAN in urban air exhibit pronounced diurnal profiles, the observed differences in activity may reflect different exposures of PAH to these oxidizing species.

Also relevant to the postulated importance of oxidative processes during atmospheric transport of PAH are several reports of the carcinogenic activity of polar fractions of organic particulates (Hueper *et al.* 1962; Epstein, Mantel & Stanley 1968; Wynder & Hoffman 1965; Asashina *et al.* 1972; Gordon *et al.* 1973), products of ozonized gasolines (Kotin, Falk & McCammon 1958) and products of oxidation of aliphatic hydrocarbons (Kotin, Falk & Thomas 1956); and of the toxicity of the photooxidation products of a commercial fuel oil (Larson, Hunt & Blankenship 1977).

There are also several reports that BaP and other PAH readily undergo photochemical transformations when adsorbed on a variety of support materials such as filters, silica gel and carbon (soot) particles (Kotin *et al.* 1954; National Academy of Sciences 1972; Falk, Markul & Kotin 1956; Tebbens, Thomas & Mukai 1966; Tebbens, Mukai & Thomas 1971; Boyer & Laitinen 1975; Barofsky & Baum 1976).

Finally, it has been stated that polycyclic quinones are present in ambient particulates and that they may be formed by atmospheric photooxidation of PAH (Masuda & Kuratsune 1966; Fatiadi 1967; Jaeger 1971; Pierce & Katz 1976).

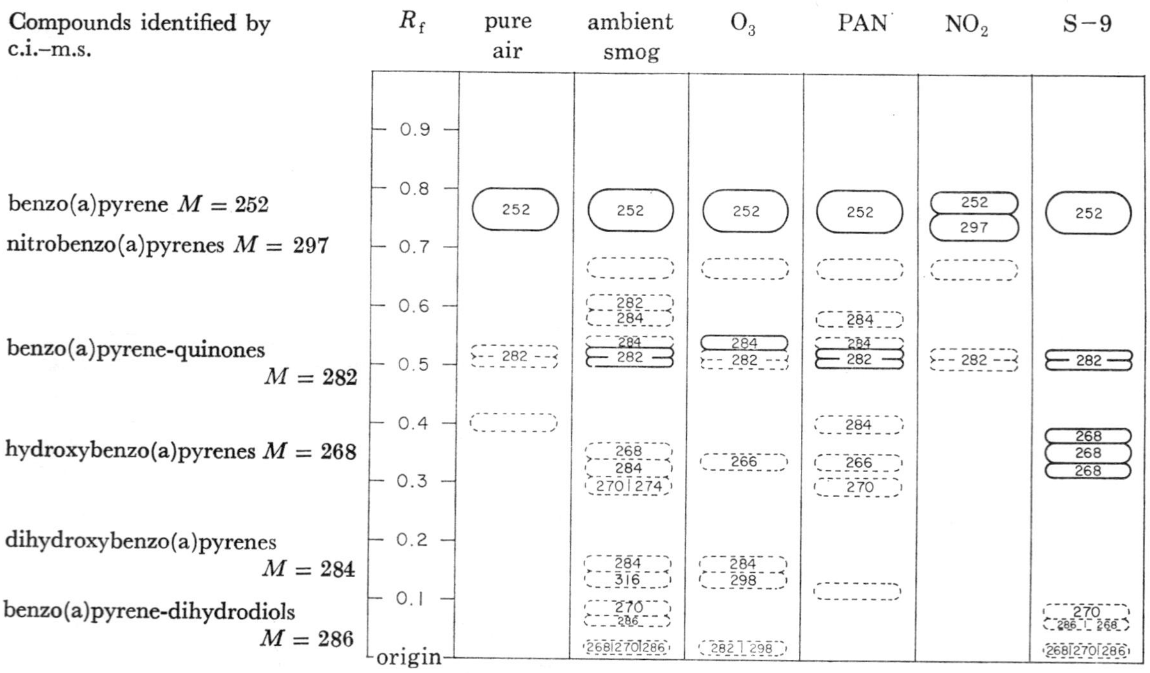

FIGURE 1. Thin layer chromatography separation of the reaction mixtures from BaP exposed to several gaseous pollutants, to ambient photochemical smog and to pure air. For reference, t.l.c. bands are also shown of metabolites formed from the action of liver homogenate (S-9) on BaP. The solvent was toluene:dichloromethane:methanol (25:10:1).

2.4.1. *Reactions of benzo(a)pyrene*

In order to test our hypothesis, a series of experiments was carried out in both real and simulated atmospheres. The first question to be answered was 'Can BaP deposited on a filter react with the gases present in ambient photochemical smog to form compounds which, unlike BaP, are direct mutagens?'

Experimentally, two washed and fired Gelman A/E glass fibre filters were mounted in series in a high-volume sampler capable of drawing 1130 l/min of ambient air through them. The upstream filter, a 'blank', simply collected all ambient particulates and allowed only the gaseous pollutants present in the photochemical smog to pass through to the second. This filter was coated with BaP (*ca.* 1 mg) that could interact with the gaseous pollutants. In these and all other runs, the filters coated with PAH were kept in the dark throughout the exposure. In a separate 'blank' run without BaP no transfer of mutagenic species from the first filter to the second was observed.

Four such runs were carried out, two in Los Angeles and two in Riverside. At each site one test was conducted during the day and one at night; total exposures to ambient air were about 40 h per sample.

After exposure, the filter samples were extracted by ultrasonication and the concentrated organic extracts tested with *Salmonella* strains TA98 and TA100. All samples were directly mutagenic; typically with TA98, 10 μl of extract produced 150–250 revertants per plate *without* metabolic activation. Thus exposure of BaP in the dark to the gases in ambient smog forms compounds that are directly active mutagens.

In order to determine the chemical nature of these species, a similar experiment was performed in which ambient Riverside air was drawn through the pair of filters for 72 h. The average concentrations during this period for O_3, PAN, NO_2 and CO were 0.035, 0.0025, 0.050 and 1.10 parts/10^6 respectively.

The BaP-coated filter was then extracted and separated into fractions by thin-layer chromatography (t.l.c.) on silica gel plates. Each band was recovered in methanol and analysed by methane chemical ionization mass spectrometry. The t.l.c. bands and the molecular masses of the species formed are shown in the column headed 'ambient smog' in figure 1. Also shown at the left of the figure are the products identified by direct probe mass spectrometry and the t.l.c. R_f values (solvent toluene–dichloromethane–methanol, 25:10:1).

Clearly, not only does exposure of a sample of BaP to ambient smog cause it to become directly mutagenic, but contrary to some reports (Berry & Lehman 1971; Fishbein 1976), a variety of compounds are readily formed. Those tentatively identified by t.l.c. include, in order of decreasing polarity: BaP-dihydrodiol(s) (molec. mass 286), BaP-diphenol(s) (284), BaP-phenol(s) (268), BaP-quinones (282), and what appear to be dicarbonyl compounds (284) formed by ring-opening oxidations.

The recovery of pure polar compounds from t.l.c. plates is somewhat inefficient. Furthermore, the treatment of the BaP–smog products during extraction, separation and analysis was rather severe and time consuming so that some initial products may have reacted before their final identification. Thus, although no mass spectral evidence for BaP-epoxides was obtained, the presence of dihydrodiol(s) suggests epoxides may have been formed as initial, labile products. Detailed experiments using h.p.l.c. are currently under way to resolve this matter.

Based on elementary carbon analysis one obtains a rough estimate of the conversion of the BaP on the filter to products, which indicates that the surface area of the BaP directly exposed to the polluted air is the factor limiting the degree of oxidation in experiments of this type. This is consistent with the observations of Lane & Katz on the ozonization of BaP (Lane 1976; Lane & Katz 1977).

To obtain more specific information about the reactions of BaP with individual oxidants present in photochemical smog, another series of experiments was carried out under similar conditions. In these, glass-fibre filters coated as above with BaP were exposed in the dark to clean, particle-free air containing (1) 11 parts $O_3/10^6$ (exposure time 24 h at a flow rate of 85 l/min), (2) 1.3 parts $NO_2/10^6$ containing 10 parts nitric acid/10^9 (24 h, 28 l/min), or (3) 1.1 parts PAN/10^6 (16 h, 85 l/min). Control runs with BaP exposed to pure air (24 h at 85 l/min) and with blank filters exposed to NO_2, O_3, or PAN were also included. No mutagenicity was observed in any of the control runs but all the other exposures resulted in direct mutagenicity of the reaction mixtures.

After exposure, the products and unreacted BaP were separated by t.l.c., analysed by mass spectrometry and tested separately for mutagenic activity by the Ames test, as described earlier.

Microsomal activation was achieved by adding 0.5 ml of liver S-9 solution from Aroclor-induced rats. In a separate experiment, for product identification and comparison purposes, a sample of 9 mg of BaP was incubated for 30 min at 37 °C with the liver S-9 homogenate and the metabolites formed in this microsomal activation system analysed and tested for mutagenic activity.

It is evident from the results in figure 1 that BaP reacted with O_3 and PAN to give several oxidized products; with NO_2 and a trace of nitric acid it gave almost exclusively mononitrobenzo(a)pyrene(s). BaP reacted to only a slight degree with pure air, but the mixture was non-mutagenic without activation.

In recent experiments, the exposure levels of BaP on glass fibre filters were lowered to about 100 parts $O_3/10^9$ and 250 parts $NO_2/10^9$; similar results were observed. Conversion yields measured by fluorimetry were *ca.* 20 % for the nitration (8 h exposure) and *ca.* 47 % for the reaction with ozone (1 h exposure).

The nitration of BaP by levels of NO_2 of the order of 1 part/10^6 in air is catalysed by levels of nitric acid or sulphuric acid around 1 part/10^9. This species was shown by long-path (100 m) Fourier transform infrared spectroscopy to be present in our tanks containing known amounts of reportedly 'pure NO_2 in nitrogen.'

The t.l.c. bands containing the unreacted BaP (R_f = 0.77) were not directly active as expected and required metabolic activation to produce mutagenic effects. The bands containing the BaP-quinones (R_f = 0.54) were complex; they contained, in addition to quinones, trace amounts of an unidentified directly mutagenic compound. Authentic samples of the three purified quinones (BaP-6,12-quinone, BaP-1,3-quinone and BaP-3,6-quinone) were tested in our laboratory and were found to be non-mutagenic in the Ames assay system. However, caution should be expressed in the interpretation of t.l.c. separation patterns in terms of the formation of BaP-quinones in the exposures of BaP both to ambient air and to ozone. Indeed, recent separation of the reaction mixture of BaP with O_3 by h.p.l.c. techniques did not reveal the presence of any quinones in the reaction mixture. Obviously, at least some formation, and therefore detection, of quinones in the t.l.c. separation of BaP-containing mixtures may be attributed to photochemical transformation of the BaP adsorbed on the t.l.c. plate.

Treatment of BaP with liver S-9 mix resulted in the appearance of a complex t.l.c. band (figure 1) containing directly active mutagens. Its R_f (0.35) and the molecular mass (268) of its components are consistent with those of isomers of hydroxybenzo(a)pyrene.

Exposure of BaP to *ca.* 1 part $NO_2/10^6$ (and a trace of HNO_3) in air resulted in the appearance of only one major t.l.c. band; it was directly mutagenic. Its R_f (0.74) and molecular mass (297) are consistent with the structure mononitrobenzo(a)pyrene (nitro-BaP). By changing from a mixed t.l.c-solvent system to toluene alone, this band was resolved into two bands, one yellow and one orange. The latter had a lower R_f value. By comparison of ultraviolet-visible spectra shown in figure 2 and their mass spectra with those of authentic samples synthesized by the method of Dewar, Mole, Urch & Warford (1956*a*), we assigned the structure 6-nitro-BaP to the compound in the yellow t.l.c. band; the orange t.l.c. band consisted of a mixture of the 1-nitro and 3-nitro isomers.

Co-chromatography of the two t.l.c. bands with authentic samples further confirmed these assignments. The two isomers present in the orange t.l.c. band were then resolved by gas chromatography with the use of a glass capillary column. Interestingly, the relative concentrations of the 1-nitro and 3-nitro isomers were 1:1 in the mixture synthesized according to

Dewar *et al.* (1956*a*) and 10:1 in the orange t.l.c. band formed upon exposure of BaP to NO_2 in air.

Both 6-nitro-BaP and the mixture of the 1-nitro and 3-nitro-BaP isomers proved to be direct frameshift mutagens when tested with strains TA1537, TA1538, TA98 and TA100. The 6-nitro-isomer was a moderate mutagen (32 rev./nmol for TA100) while the mixture of 1- and 3-isomers showed strong mutagenic activity (210 rev./nmol for TA98). Addition of liver S-9 homogenate significantly enhanced the mutagenic activity observed with strains TA1537, TA1538 and TA98. Thus, with S-9 the 6-nitro-BaP caused 418 rev./nmol (TA100) and the mixture of 1- and 3-nitro 5200 rev./nmol (TA98). However, the addition of S-9 actually reduced the activity of the mixture of 1-nitro and 3-nitro-BaP on TA100. Dose–response curves for the mutagenic activity of the authentic samples were in excellent agreement with those of the nitro-BaP isomers formed upon exposure of BaP to NO_2 in air. No base-pair substitution activity was found when the samples were tested with strain TA1535.

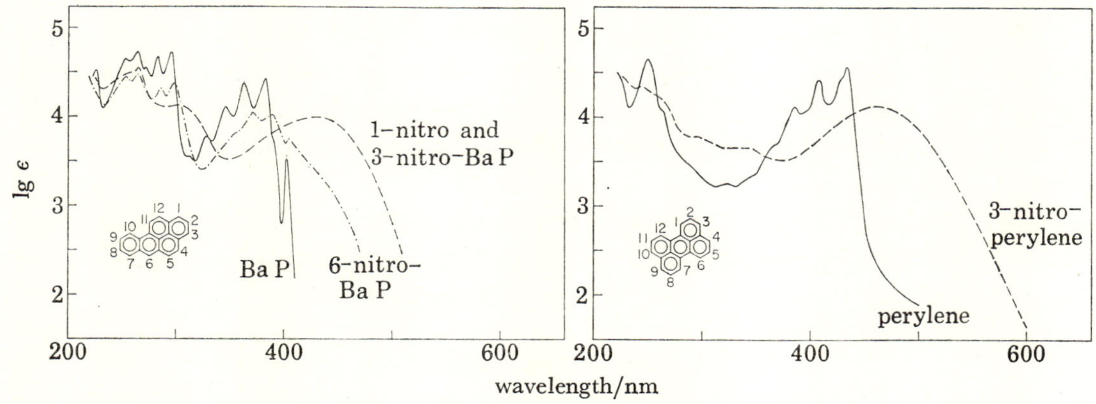

FIGURE 2. U.v. – visible spectra in methanol of the nitro derivatives of BaP and perylene formed upon exposure of the PAH deposited on a filter to 1 part $NO_2/10^6$ (and a trace of HNO_3) in air.

We can conclude from these observations that the pattern of compounds resulting from the exposure of BaP to ambient photochemical smog can be reconstituted in part by combining the individual effects of ozone and PAN on the BaP (figure 1). Furthermore, it is seen that metabolities formed by treating BaP with S-9 rat liver homogenate coincide both in molecular mass and mutagenicity, with several t.l.c. fractions observed when BaP was treated with ambient photochemical smog.

Several factors may explain the fact that nitro-BaP was not found in the experiment when BaP was exposed to ambient smog. Thus, NO_2 levels are customarily low in Riverside and for the 72 h exposure averaged only 0.050 part/10^6. It is also possible that the prolonged exposure of 72 h to other oxidizing species present in the smog resulted in the degradation of any nitro-BaP compounds that may have been formed initially. Recently, 6-nitro-BaP was detected in the air in Prague (Jäger 1978).

2.4.2. *Reactions of non-mutagens perylene and pyrene*

Since exposure of a known carcinogen and activatable mutagen BaP to only 1 part $NO_2/10^6$ in air produced directly mutagenic mononitro derivatives, we were interested to see if under similar conditions the same type of reaction could take place with a non-mutagenic (in the Ames reversion assay) PAH. To answer this question perylene, a closely related isomer of

BaP that is also found in organic particulates from petroleum and diesel engine exhaust, tobacco smoke, effluent from coal-fired furnaces, etc., was chosen. It was deposited on glass-fibre filters and, as with BaP, exposed 24 h in the dark to 1 part $NO_2/10^6$ (and a trace of HNO_3) in air at a flow rate of 28 l/min.

Separation of the exposed sample by t.l.c. yielded only one reaction product band. It was brick red, and from its mass spectrum and u.v. spectrum (figure 2), the compound responsible was identified as 3-nitroperylene. When tested with strains TA98 and TA1538, 3-nitroperylene was found to be a directly active, moderate mutagen (37 rev./nmol, TA1538) whose activity was significantly enhanced by addition of liver S-9 mix (94 rev./nmol, TA1538).

To check the generality of this transformation, i.e. conversion of a non-mutagenic (in the conventional Ames test), non-carcinogenic PAH to a directly active species, we recently deposited the quadricyclic PAH pyrene on a glass-fibre filter and exposed it to *ca.* 1 part $NO_2/10^6$ (and a trace of HNO_3) in air under similar conditions. Indeed, 1-nitropyrene was formed, though, as expected from theoretical considerations on aromatic nitrations in solution (Dewar, Mole & Warford 1956*b*), at a lower rate and lower yield than 6-nitro-BaP, 1- and 3-nitro-BaP, or 3-nitro-perylene. 1-Nitropyrene proved to be a direct mutagen (TA1538, *ca.* 100 rev./nmol) whose activity was in fact reduced when S-9 was added. Such S-9 suppression is quite common and the possibility must be kept in mind when carrying out metabolic activation.

In a similar experiment chrysene, an isomer of pyrene, was *not* nitrated. This seems reasonable since the relative rates for aromatic nitration in solution are: chrysene = 1, pyrene = 4.9, perylene = 22 and BaP (6-position) = 31 (Dewar *et al.* 1956*b*).

2.5. *Photochemical reactions*

The fact that upon photolysis 9-nitroanthracene forms 9,10-anthraquinone, both in solution (Chapman, Heckert, Reasoner & Thackaberry 1966) and in recent experiments on silica gel (Pitts *et al.* 1978*c*), suggests that similar reactions may represent, at least in part, the fate of nitro-BaP (and possibly other PAH) in ambient atmospheres. Thus, we feel that the photo-oxidation of PAH by sunlight in air (Falk *et al.* 1956; Tebbens *et al.* 1966; Tebbens *et al.* 1971; Boyer & Laitinen 1975), for the formation of polycyclic quinones found in polluted air, e.g. in Toronto, Canada (Pierce & Katz 1976). Indeed, recent experiments in our laboratory have shown that irradiation of 6-nitro-BaP deposited on silica gel yields the expected quinones.

At present one can only speculate whether or not the mechanism proposed by Chapman *et al.* (1966) for the formation of 9,10-anthraquinone as a major product of the irradiation of 9-nitroanthracene would also apply to 6-nitro-BaP.

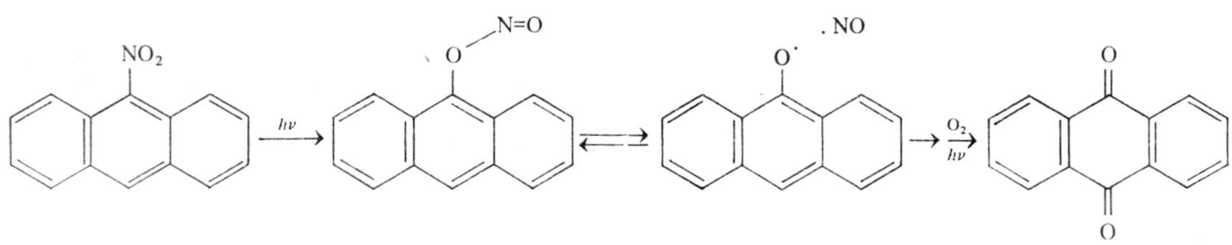

Certainly, the electron impact mass spectra of the nitro derivatives of BaP and perylene are similar in some respects to that of 9-nitroanthracene. Thus loss of 30 u, corresponding to NO, is an important process in all cases. Presumably, on electron impact the nitro group rearranges to a nitrite with subsequent loss of the NO. This process is analogous to the photorearrangement postulated by Chapman *et al.* (1966) for 9-nitroanthracene. Such parallels between intramolecular rearrangements observed photochemically and by electron impact in mass spectrometers are well known (Calvert & Pitts 1966) and support the hypothesis that 6-nitro-BaP may photooxidize in smog to form the quinones. Experiments to test this idea are under way in our laboratory.

2.6. *Environmental implications*

The formation of mutagenic nitro derivatives by exposure of non-carcinogenic as well as carcinogenic PAH to levels of NO_2 around 1 part/10^6 warrants further investigations of the modes of formation and atmospheric fate of these compounds, especially in situations where relatively high levels of PAH and oxides of nitrogen may coexist and direct mutagens may be formed. These could include spark ignition (Wang, Sawyer & Wei 1978) and diesel exhausts (Huisingh *et al.* 1978) as well as plumes from coal-fired power plants (Chrisp, Fisher & Lammert 1978).

The significant reactivity of BaP and perylene (and even pyrene), we observed with NO_2 in the dark, as well as BaP with O_3, and PAN and ambient photochemical smog, strongly suggests that similar processes could occur in ambient atmospheres – possibly to a major degree. Furthermore, these reactions with BaP do not constitute unique pathways for the formation of such oxidized species as the quinones. Thus at least two other mechanisms can be invoked for the atmospheric photodegradation of PAH.

One path would be electrophilic attack on the PAH by singlet molecular oxygen, $O_2(^1\Delta)$, presumably to produce in part hydroperoxides. These could then decompose to phenoxy-type radicals, which in the case of BaP would be converted to BaP-quinones (Inomata & Nagata 1970). Atmospheric sources of $O_2(^1\Delta)$ in sunlight include triplet energy transfer from photoexcited organic species such as PAH or benzaldehyde to ground state molecular oxygen (Pitts, Khan, Smith & Wayne 1969). Evidence for this mechanism has been obtained recently in experiments in which BaP coated on particles suspended in air was oxidized to quinones by $O_2(^1\Delta)$. The latter was produced by energy transfer from particles coated with rose bengal and illuminated at 365 nm (Grossman 1976; Grossman & Van Cauwenberghe 1976, private communication).

Other highly reactive species present in photochemical smog include hydroxyl and hydroperoxyl radicals, both of which would undoubtedly attack PAH. Not much is known about HO_2, but certainly the reactions of OH with gas phase aromatics are very fast, greater than $3 \times 10^9\ l\ mol^{-1}\ s^{-1}$ (Perry, Atkinson & Pitts 1977; Atkinson *et al.* 1978*a*), and would probably lead to BaP-phenols and ultimately BaP-quinones.

Finally, we should emphasize that our studies were conducted with PAH deposited on the surface of *glass fibre filters*. Whether PAH adsorbed on the surface of airborne particles (soot, fly ash, etc.) will react with co-pollutants in a similar fashion *in the atmosphere* is currently a matter of conjecture. Thus, the atmospheric reactions may be influenced not only by the levels of PAH and gaseous pollutants but also by many complex factors typical of surface chemistry and aerometry including particle size, sunlight intensity, relative humidity, atmospheric mixing and transport time, etc. Similarly, little is known about the extent of possible reactions of PAH on the filters employed to collect ambient particulates. However, our results suggest that they may indeed be significant.

Therefore, the elucidation of possible filter 'artefacts' is important because virtually all previous studies of the carcinogenic and mutagenic activity of organic particulates present in polluted air and motor exhaust have been based upon samples collected on various types of filters.

3. Atmospheric chemistry of amines

In the recent E.P.A. document ('S.T.A.R.') on nitrosamines in the environment (U.S. E.P.A. 1976) the following statement is made: 'In the face of these uncertainties, the existing information on the atmospheric chemistry of nitrosamines can have only vague implications regarding the control of atmospheric nitrosamines. Thus, to decide whether the nitrosamine problem is an ambient-atmosphere problem or an industrial-atmosphere problem requires that the nitrosamine to precursor dependencies be better known.'

We shall now discuss current results bearing on the problems cited above: the reactions of amines and NO_x in real and simulated urban atmospheres. In doing so, examples will be given of situations in which injudicious extrapolation of results on photochemical processes and reaction mechanisms obtained under 'laboratory' conditions (i.e. ranging from pressures around 1 Torr to concentrations around 1 part/10^6) to predict the behaviour of amines and NO_x in the parts/10^6 to parts/10^9 concentration range typical of actual ambient atmospheres can lead to erroneous conclusions.

The uncertainties inherent to such kinetic and mechanistic extrapolations – while far from the magnitude of those toxicologists often are required to make, predicting from animal data health effects of environmental pollutants on man – nevertheless can be substantial. Furthermore, they can have pronounced effects on the economic as well as technical impacts of air pollution control strategies. Potential errors can be reduced by employing several experimental systems capable of covering concentrations ranging from parts/10^3 to parts/10^9 of the pollutants, their precursors and their reaction products. This approach will be illustrated in the following discussion.

3.1. *Chemical inhibition of photochemical smog*

Recently, a proposal for controlling the products of photochemical smog such as O_3, peroxyacetyl nitrate (PAN), secondary aerosols, etc., by adding to urban air 100 parts/10^9 of an inhibitor, diethylhydroxylamine (DEHA), $(C_2H_5)_2NOH$, has been strongly advocated (Heicklen 1976). This chemical approach, it is stated, would eliminate the need for exhaust control devices on motor vehicles, and save billions of dollars. Actual field trials involving introduction of DEHA into the air of several major cities in the world suffering from photochemical smog have been suggested.

The research on which the proposal was originally based was carried out on hydrocarbon–NO_x systems containing DEHA at concentration levels over 1 part/10^6 and generally conducted under 'laboratory', not simulated ambient conditions. Thus our group, among others (Maugh 1976), was concerned not only with the possible health effects of photooxidation products of DEHA–NO_x–HC mixtures generated in ambient photochemical smog, but also with the atmospheric uncertainties involved in this extrapolation. It seemed unlikely that such low levels of an amine (100 parts/10^9) could continually and effectively inhibit ozone, PAN and secondary aerosol formation as a smog front travelled in some cases hundreds of kilometres over a period of many hours, and even days.

To test the idea under realistic conditions, we studied the reactions of DEHA at 100 and 500 parts/10^9 in ambient air, or in hydrocarbon surrogate mixtures simulating 6 to 9 a.m.

Los Angeles air (Pitts, Smith, Fitz & Grosjean 1977*b*), in a large outdoor environmental chamber constructed from FEP Teflon film 50 μm thick.

Typically, the chamber was first covered with a black plastic light shield and filled with ambient air to a volume of about 50 m^3. It was then divided into two compartments of equal volume and a known amount of DEHA in high-purity N_2 added to one of the compartments. The black cover was then removed, exposing both compartments to sunlight for at least 6 h. Since identical samples of polluted air, one with and one without DEHA, were subjected simultaneously to the same environmental conditions (i.e. solar irradiation, temperature and humidity), differences in the time–concentration profiles of the smog parameters measured in the two compartments could unambiguously be attributed to the DEHA in one compartment.

Results from an experiment in which DEHA at 100 parts/10^9 was present in Riverside ambient air are shown in figure 3. Clearly, the DEHA did *not* inhibit formation of photochemical smog; in fact, it substantially *accelerated* the conversion of NO to NO_2 and the production of O_3, PAN and light-scattering secondary aerosols.

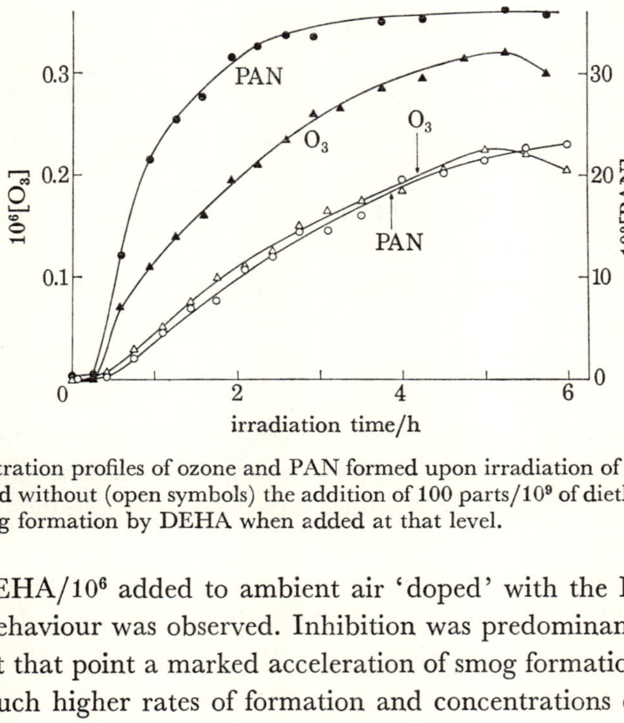

FIGURE 3. Time–concentration profiles of ozone and PAN formed upon irradiation of Riverside ambient air with (closed symbols) and without (open symbols) the addition of 100 parts/10^9 of diethylhydroxylamine, showing acceleration of smog formation by DEHA when added at that level.

With 0.5 parts DEHA/10^6 added to ambient air 'doped' with the HC surrogate mixture, an 'intermediate' behaviour was observed. Inhibition was predominant during the first hour of irradiation, but at that point a marked acceleration of smog formation took place. This was evident from the much higher rates of formation and concentrations of O_3, PAN and light-scattering particles in the compartment with DEHA.

In short, DEHA is indeed an inhibitor of certain symptoms of photochemical smog for several hours when present at concentrations around 1 part/10^6. However, it is an *accelerator* of O_3, PAN, aerosol formation and NO–NO_2 conversion when present at the recommended level of 100 parts/10^9. Furthermore, even if higher levels of DEHA (odour threshold 0.5 parts/10^6) could be added to ambient air (not considering the health implications, which could be serious), its dilution as the air mass was being transported would soon lead to *increased* photochemical smog in suburban, and even rural areas downwind.

Clearly, proposals for the control of photochemical smog by chemical additives cannot be based only on extrapolations from experiments conducted in the 'laboratory' torr pressure to parts/10^9 concentration ranges. They must be experimentally validated at realistic pollutant levels in ambient air.

3.2. *Formation of nitrosamines and nitramines*

In the gas phase, even under 'laboratory' conditions in the torr pressure range, relatively little has been published on the chemistry of amine–NO_x–HONO–air systems. The most definitive work is that of Hanst, Spence & Miller (1977), who recently reported that 1 part/10^6 of gaseous dimethylamine (DMA) reacts in air at a rate of *ca.* 4 %/min with 0.5 part/10^6 of gaseous HONO (in equilibrium with 2 parts/10^6 NO, 2 parts/10^6 NO_2, and 13000 parts/10^6 H_2O) to produce dimethylnitrosamine (DMN) with an approximate 10–30 % yield. They also showed that DMN is rapidly photolysed in sunlight and that a new, 'unknown' compound is a major product.

We have used two quite different smog chambers and analytical techniques such as gas chromatography – mass spectrometry (g.c.–m.s.) and long-path Fourier-transform infrared (F.t.–i.r.) spectroscopy, to answer such questions as:

Can nitrosamines be formed when levels of secondary alkylamines below 1 part/10^6 are introduced into air containing typical ambient levels of NO_x and small amounts of HONO?

Do tertiary aliphatic amines also react with NO_x and HONO under similar conditions to form nitrosamines?

What are the rates of conversion of the amines, and what are their products and yields?

What is the relative importance of thermal (dark) versus photochemical processes in nitrosamine formation and decay?

In the range below 1 part/10^6, do simple aliphatic amines inhibit or accelerate formation of photochemical oxidant?

Are compounds other than nitrosamines being formed that may also represent possible environmental hazards?

3.2.1. *Experimental*

Two outdoor environmental chambers were employed, one similar to that used in the DEHA experiments and the other designed for spectroscopic studies. Each allowed the study of simulated polluted atmospheres under realistic conditions of solar irradiation (May–June 1977, latitude 33° N), temperature and humidity.

In a typical experiment, the large 50 m^3 chamber (not subdivided in these experiments) was initially covered with black plastic film and filled with particle-free air at 20–40 % relative humidity (r.h.) from our air purification system (Doyle, Bekowies, Winer & Pitts 1977). A measured volume of NO_2 was injected into a dilution flask and flushed into the chamber with 30 l O_2; NO in N_2 was injected in the same way 3 min later and the contents of the chamber thoroughly mixed. Under these conditions, some nitrous acid was formed, but its concentration could not be determined. Samples of diethylamine (DEA) or triethylamine (TEA) (vacuum-distilled, purity-verified by g.c.–m.s. and i.r.) were injected 15 min later. Initial amine concentrations, as determined by g.c. with the use of Tenax as a preconcentration step, were *ca.* 0.45 part/10^6 for DEA and 0.35 part/10^6 for TEA.

The amine–NO_x–HONO mixtures were allowed to react for 2 h in the dark. The black cover was then removed and the chamber exposed to sunlight for another 2 h. Ozone, NO, NO_2, PAN, acetaldehyde and the aerosol light-scattering coefficient (b_{scat}) were continuously monitored. Gas samples for g.c.–m.s. analysis were collected at regular intervals on Tenax g.c. cartridges and analysed on a Finnigan Model 3100 instrument coupled to a Model 6100 mass spectral data system. Concentrations of DEA and TEA were monitored. Small (less than 10 %) corrections were made for artefact formation of diethylnitrosamine (DEN) on the Tenax.

The other environmental chamber contained a multiple reflexion i.r. cell having an optical path up to *ca.* 1.2 km (figure 4). Four rectangular mirrors, cut from a single 41 cm diameter Pyrex blank, composed the in-focus end of this cell, originally developed by Hanst and associates of the E.P.A. and modified in our laboratory. The out-of-focus optical assembly consisted of four 30 cm diameter mirrors set 22.5 m (the average radius of curvature of the eight mirrors) from the nesting mirror assembly. All mirrors were optically polished and gold coated for maximum reflectivity in the infrared.

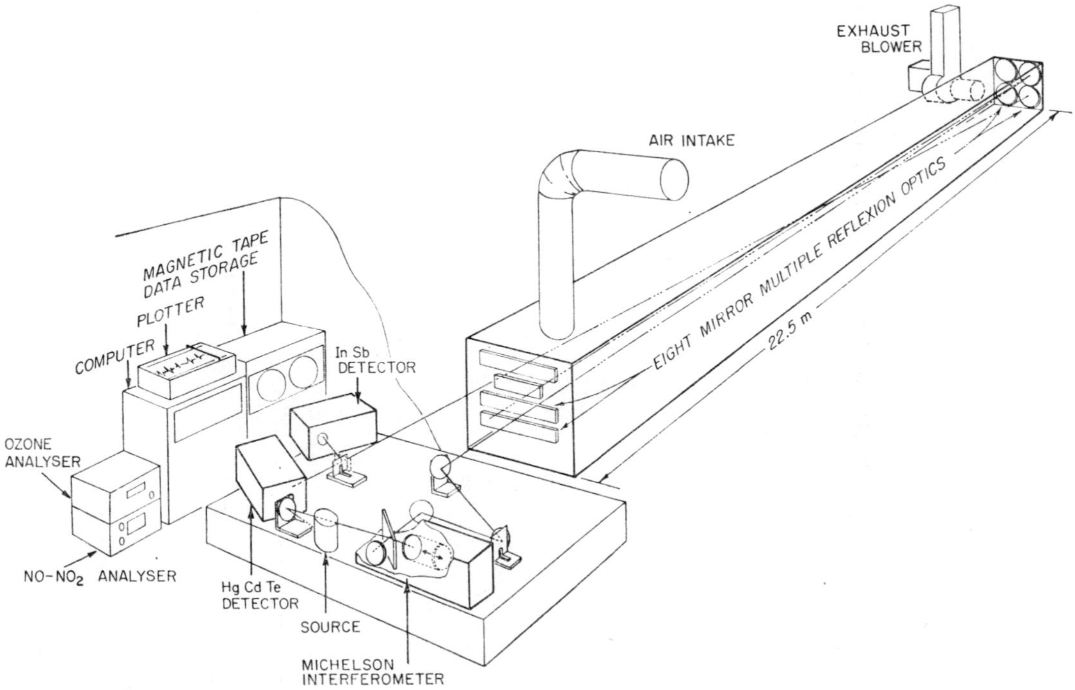

FIGURE 4. One kilometre optical pathlength Fourier-transform infrared spectroscopic facility.

The cell housing (i.e. the chamber) consisted of a rectangular aluminum frame (0.76 × 0.76 × 23 m) enclosed by 50 μm FEP Teflon walls whose measured transmittance was more than 80 % in the actinic u.v. For dark reactions a removable sun screen could be installed over the entire length of the cell.

Coupled with the cell was a Digilab Model 296 Michelson i.r. interferometer with a variable spectral resolution up to 0.5 cm^{-1}. The beam leaving the cell was sent to either of two liquid N_2 cooled detectors: a photovoltaic InSb detector for the 2000–3900 cm^{-1} region or a photoconductive HgCdTe detector for the 600–2000 cm^{-1} region. Data collection and processing were performed by a Data General Nova 1200 computer (Tuazon, Winer, Graham & Pitts 1977; Tuazon *et al.* 1978*a*, *b*). Initial concentrations employed with the two experimental systems were *ca.* 0.1–0.2 part $NO_x/10^6$ and *ca.* 0.3–0.5 part amine/10^6 in the 50 m^3 chamber and *ca.* 4 parts/10^6 NO, NO_x and amine in the F.t.–i.r. cell (r.h. *ca.* 40–50 %).

3.2.2. *Results*

(*a*) *Large outdoor chamber*. Solar irradiation of diethylamine-NO_x-air and triethylamine–NO_x–air mixtures at concentrations below 1 part/10^6 in the 50 m^3 chamber resulted in typical manifestations of severe photochemical smog, i.e. rapid conversion of NO to NO_2 and the formation of substantial amounts of O_3, PAN, CH_3CHO and light-scattering particles. Time–concentration profiles for DEA are shown in figure 5. Clearly, at these concentrations, both DEA and TEA behave as organics of high reactivity in producing photochemical smog.

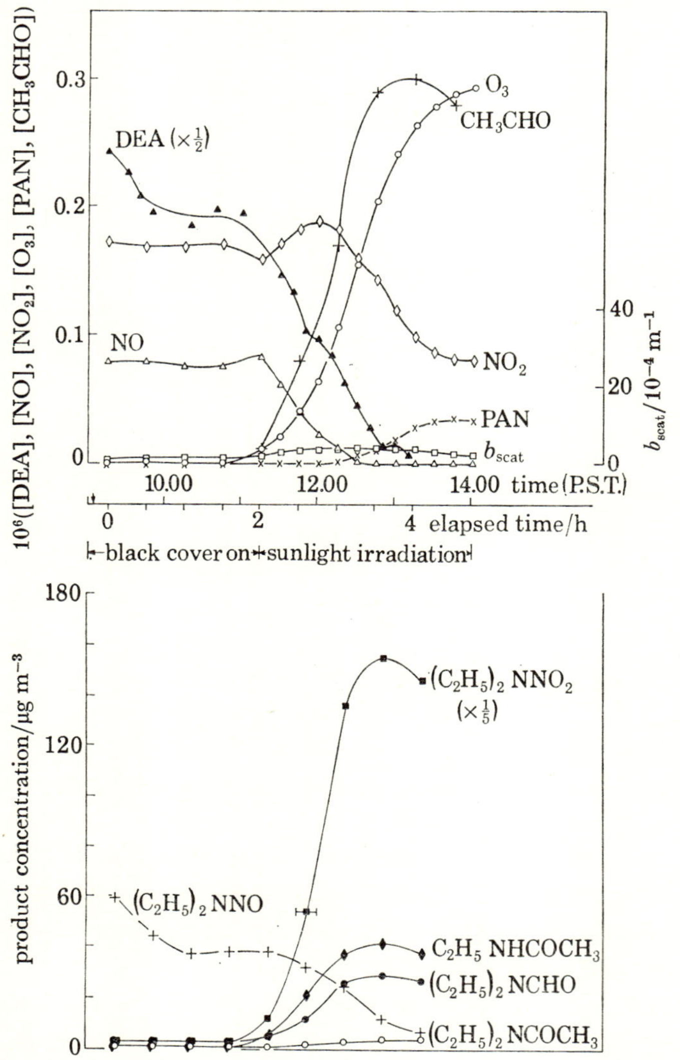

FIGURE 5. Time–concentration profiles of diethylamine, NO, NO_2, PAN, acetaldehyde and other nitrogenous compounds in the diethylamine–NO_x–HONO–air system under simulated atmospheric conditions in the dark and in sunlight.

As seen in figure 5, diethylnitrosamine was rapidly formed in the dark from the secondary amine DEA, but with a rather low yield of 59 μg/m^3 (maximum = 2.8 %). It was destroyed in sunlight with $t_{\frac{1}{2}} \approx 1$ h. A small amount of DEN was also formed in the dark from TEA (yield *ca.* 0.8 %). However, in contrast to DEA, on irradiation of the TEA–NO_x mixture with sunlight, *additional* DEN was formed initially with a maximum yield of 38 μg/m^3 (*ca.* 1.8 %).

Clearly, the idea that dialkylnitrosamines cannot be formed in sunlight from simple tertiary amines and NO_x is incorrect.

Four major gas phase products (other than acetaldehyde and PAN) were formed upon irradiation of both DEA–NO_x and TEA–NO_x mixtures: diethylnitramine [$(C_2H_5)_2NNO_2$], *N,N*-diethylformamide [$(C_2H_5)_2NCHO$], *N*-ethylacetamide [$C_2H_5NHCOCH_3$] and *N,N*-diethylacetamide [$(C_2H_5)_2NCOCH_3$]. Subsequently, an authentic sample of diethylnitramine was prepared and further confirmation was obtained by g.c.–m.s. and i.r. spectroscopy.

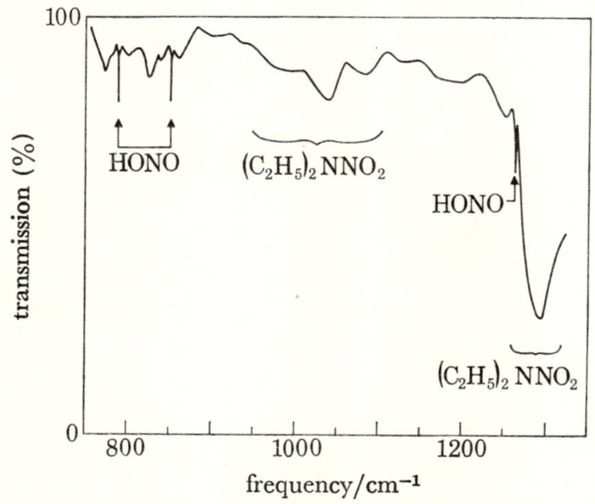

FIGURE 6. I.r. spectrum of diethylnitramine found upon irradiation of a diethylamine–NO_x–HONO mixture in air; optical pathlength 720 m.

Diethylnitramine was the major product in the DEA–NO_x system, reaching a maximum concentration of 162 parts/10^9 (780 μg m^{-3}). This corresponds to a yield of 32 % of the measured initial DEA concentration (0.45 part/10^6). The two major products from TEA–NO_x irradiations were diethylformamide (178 μg m^{-3}, 8.6 % yield), and diethylnitramine (177 μg m^{-3}, 7.4 % yield). Two additional gas phase products, not seen with DEA, were also formed in the gas phase. The major one (molecular ion $m/e = 87$) had a yield estimated at *ca.* 2 % and a mass spectrum compatible with an amide-like structure. The second product was formed in low yield and was identified as diacetamide $(CH_3CO)_2NH$.

G.c.–m.s. analysis of high volume aerosol samples collected at the end of the experiments revealed the presence of small amounts of acetamide in both the DEA run (3 μg/m^3) and the TEA run (9 μg/m^3). Diethylhydroxylamine was also formed (*ca.* 8 μg/m^3) in the TEA experiment.

(*b*) *Long-path F.t.–i.r. studies.* The F.t.–i.r. spectroscopic facility (figure 4) was employed at a path length of 720 m and a resolution of 1 cm^{-1} to investigate the infrared spectra of the products formed on irradiating with sunlight DMA–NO_x and DEA–NO_x mixtures (and associated HONO) in air at *ca.* 50 % r.h. In particular, *in situ* i.r. spectroscopic confirmation of the formation of nitramines and identification of the 'unknown' major photolysis product of dimethylnitrosamine in air observed by Hanst *et al.* (1977) were desired.

(i) *Dimethylamine*–NO_x. After several minutes of solar irradiation of DMA–NO_x mixtures ([NO] ≈ [NO_x] ≈ [DMA] ≈ 5 parts/10^6), and associated HONO, a very strong absorption band was observed at 1308 cm^{-1}. This band, as well as those appearing at 773, 985 and 1132 cm^{-1}, became more intense as the experiment progressed. Irradiations lasted for 2–4 h.

Comparison of these four major absorption frequencies with those reported by Davies & Jonathan (1958) for dimethylnitramine showed good agreement, both for the positions and relative intensities of the bands. The dimethylnitramine formed in these experiments was also identified by g.c.–m.s. analysis of samples adsorbed on Tenax cartridges. From the approximate positions of the major bands in the spectrum of the 'unknown' product of Hanst *et al.* formed in the photolysis of dimethylnitrosamine in air, we identify it as dimethylnitramine.

(ii) *Diethylamine*–NO_x. Irradiations of DEA and NO_x (and associated HONO) were carried out in the F.t.–i.r. facility in the same manner as those described above for DMA. As was the case for DMA runs, no significant amounts of O_3 or PAN were formed during these irradiations. This was due to inhibition by the high concentrations of NO and amine present; carbon monoxide and formaldehyde were, however, observed.

Contours of the product spectrum for 95 min of irradiation are shown in figure 6. For clarity, contributions from the unreacted DEA and water interferences were removed. By comparison of this spectrum with that of an authentic sample of $(C_2H_5)_2NNO_2$ prepared in our laboratory and taken under similar conditions (25°, 0.48 parts/10^6, 720 m, 1 cm^{-1} resolution), we can assign the very strong band observed at 1288 cm^{-1} to the NO_2 symmetric stretch in diethylnitramine. The absorptivity of this band is 37 cm^{-1} atm^{-1}† (Tuazon *et al.* 1978*a*).

3.2.3. *Discussion*

(*a*) *Nitrosation in the dark.* The only gaseous product identified in the dark experiments with either DEA or TEA was DEN. Its formation from DEA can be explained in terms of DEA reacting with the relatively small amounts of nitrous acid formed immediately following injection of NO and NO_2 into the 50 m^3 chamber:

$$NO + NO_2 + H_2O \underset{b}{\overset{a}{\rightleftharpoons}} 2HONO, \tag{1}$$

$$(C_2H_5)_2NH + HONO \longrightarrow (C_2H_5)_2NNO + H_2O. \tag{2}$$

The degree to which (1) and (2) occur by homogeneous versus heterogeneous paths in gas phase systems is still controversial (Chan, Nordstrom, Calvert & Shaw 1976; Kaiser & Wu 1977).

The maximum DEN yield is only 2.8 %, despite the fact that the concentrations of NO_x and DEA remain large throughout the 2 h period in the dark. One explanation is based on the fact that reaction (1) is very slow in the range 0.1–0.2 parts/10^6 (Chan *et al.* 1976; Kaiser & Wu 1977). Thus the HONO formed in the dark at the start of the experiment, when relatively high concentrations of NO and NO_2 were added consecutively to the chamber, would not significantly be regenerated after the initial amount present is consumed in (2). If this is indeed the case, the initial HONO present when the amine was injected would limit the DEN yield. Results of subsequent experiments, in which the initial HONO concentration was varied, are consistent with this hypothesis.

This is another example of the extrapolation problem in atmospheric chemistry referred to earlier. Clearly, when one considers the $NO + NO_2 + H_2O \rightleftharpoons HONO$ equilibrium in air, an assumption that the rapid rate of HONO formation observed at relatively high concentrations of reactants (e.g. 5 parts/10^6 NO and NO_2) is also true for concentrations around 10 parts/10^9 would be in serious error.

† 1 atm = 101325 Pa.

In the case of DEN formation in the dark from TEA, it is difficult to formulate a reasonable homolytic gas phase reaction mechanism. Heterolytic nitrosation mechanisms, perhaps on the chamber walls, may need to be invoked.

(*b*) *Mechanism of photo-oxidation.* The nature and distribution of the products observed for the TEA–NO_x system can be reasonably well explained in terms of initial OH attack on the secondary H atoms (reaction 3), followed by a series of reactions well established for photochemical smog (Demerjian, Kerr & Calvert 1974; Finlayson & Pitts 1976; Finlayson-Pitts & Pitts 1977):

$$(C_2H_5)_2NCH_2CH_3 + OH \longrightarrow \underset{\mathbf{1}}{(C_2H_5)_2N\dot{C}HCH_3} + H_2O. \tag{3}$$

Sources of OH radicals include the photolysis of nitrous acid at the onset of irradiation and photochemical reactions of acetaldehyde and ozone as the reaction proceeds.

By analogy with reactions of alkyl radicals in photochemical smog, one postulates that radical **1** reacts with molecular oxygen to yield the peroxy radical **2**; this then reacts with NO to yield NO_2 and the amine-alkoxy radical **3**:

$$(C_2H_5)_2N\dot{C}HCH_3 + O_2 \longrightarrow \underset{\mathbf{2}}{(C_2H_5)_2N\overset{O-O\cdot}{\overset{|}{C}}HCH_3}, \tag{4}$$

$$(C_2H_5)_2N\overset{O-O\cdot}{\overset{|}{C}}HCH_3 + NO \longrightarrow \underset{\mathbf{3}}{(C_2H_5)_2N\overset{O\cdot}{\overset{|}{C}}HCH_3} + NO_2. \tag{5}$$

Decomposition of **3** by two intramolecular pathways, (6) and (7), gives the major product acetaldehyde and the major nitrogenous product (other than PAN) diethylformamide:

$$(C_2H_5)_2N\overset{O\cdot}{\overset{|}{C}}HCH_3 \longrightarrow CH_3CHO + \overset{\mathbf{4}}{(C_2H_5)_2N\cdot}, \tag{6}$$

$$\longrightarrow (C_2H_5)_2NCHO + \cdot CH_3. \tag{7}$$

Reaction (6) is the favoured pathway. Alternatively, molecular oxygen may abstract the H atom α to the oxygen atom. This would form HO_2 and another observed product, diethylacetamide:

$$(C_2H_5)_2N\overset{O\cdot}{\overset{|}{C}}HCH_3 + O_2 \longrightarrow (C_2H_5)_2NCOCH_3 + H\dot{O}_2. \tag{8}$$

Further photo-oxidation of acetaldehyde in the presence of NO_x would lead to the formation of PAN, another major product.

Little is known about the atmospheric reactions of the diethylamino radical **4**. However, from the product distribution, and by analogy with the very slow reaction of the amino radical NH_2 with O_2 (Levine & Calvert 1977; Lesclaux, Khê, Desauzier & Soulignac 1975), one can assume that the reaction of $(C_2H_5)_2N\cdot$ with molecular oxygen is slow. Thus, presumably **4** can react efficiently with NO and NO_2, even though they are present at much lower concentrations than O_2:

$$(C_2H_5)_2N\cdot + NO \underset{h\nu \atop b}{\overset{a}{\rightleftharpoons}} (C_2H_5)_2NNO \tag{9}$$

$$\Big\downarrow \text{ direct oxidation, } h\nu \text{ and/or thermal} \tag{11}$$

$$(C_2H_5)_2N\cdot + NO_2 \underset{h\nu \atop b}{\overset{a}{\rightleftharpoons}} (C_2H_5)_2NNO_2. \tag{10}$$

The efficient photochemical formation and photostability of nitramines compared with nitrosamines is of some interest. If indeed they are formed in the competitive reactions (9) and (10), their relative concentrations will depend in part upon the ratio of NO to NO_2, the rate constants for (9*a*), (10*a*) and (11), and their relative stabilities in sunlight as determined by the photochemical reactions (9*b*) and (10*b*). The long wavelength absorption cut-offs for DEN and diethylnitramine in methanol are 390 and 310 nm respectively. Thus, even though in condensed phases nitramines react photochemically at 253.7 nm to give the corresponding nitrosamines (Suryanarayanan & Bulusu Suryanarayanana 1972), their absorption cross sections in the actinic ultraviolet ($295 < \lambda < 400$ nm) are low. Thus their loss by direct photodecomposition (10*b*) is expected to be correspondingly small. Nitrosamines, however, absorb over a wide range of actinic u.v. and photodecompose efficiently (Hanst *et al.* 1977; Bamford 1939).

Thus, the observed relative yields of DEN versus diethylnitramine, and their time concentration profiles (figure 5) are consistent with their expected stabilities in sunlight. Furthermore, our mechanism is consistent with the results of Hanst *et al.* (1977), who, as noted above, found that photolysis of DMN gave large yields of a 'new compound' which we have identified as dimethylnitramine. The relative importance of (9*a*), (9*b*), (10*a*), (10*b*) and (11) remains to be determined, as do many other aspects of amine photo-oxidation in air; so our mechanism is clearly tentative.

For diethylamine, an analogous sequence of photo-oxidation reactions involving initial OH attack on the secondary C—H bond, would give the products acetaldehyde and ethylacetamide. However, the large yield of diethylnitramine from DEA (*ca.* 4.5 times more than from TEA) indicates that additional efficient reactions also produce the diethylamino radical, **4**. A likely candidate is initial OH attack on the N—H bond (reaction 12), which would compete with attack on the secondary C—H bonds (reaction 3):

$$(C_2H_5)_2NH + OH \longrightarrow (C_2H_5)_2N\cdot + H_2O. \tag{12}$$

Indeed, recent measurements of the absolute rate constants for OH attack on dimethylamine and trimethylamine, 6.5×10^{-11} and 6.1×10^{-11} cm^3 $molecule^{-1}$ s^{-1} respectively (Atkinson, Perry & Pitts 1978*b*) suggest that in secondary alkylamines, H atom abstraction by OH from the N—H bond is competitive with H atom abstraction from the secondary C—H bonds.

(*c*) *Environmental implications.* As we noted earlier, diethylnitrosamine and dimethylnitrosamine are potent carcinogens (Fahmy & Fahmy 1976; Walters 1977). Both dimethylnitramine (Goodall & Kennedy 1976; Druckrey, Preussman, Schmahl & Muller 1961) and acetamide (Jackson & Dessau 1961; Weisburger, Yamamoto, Glass & Frankel 1969) also show carcinogenic activity in test animals, though with less potency. The possible carcinogenicity in experimental animals of diethylnitramine and of the substituted amides identified in this study has not apparently been determined.

Amines are emitted into industrial atmospheres as a result of a wide variety of activities; some of these are secondary and tertiary aliphatic amines or related compounds such as ethanolamines. Little is known about their concentrations in ambient air, but probably they are usually very small. Thus, even though levels of NO_x and HONO from mobile and stationary sources may be appreciable (Chan *et al.* 1976), the risk of forming significant amounts of nitrosamines or nitramines in air to which the general public is exposed seems correspondingly small. However, in industrial situations in which amines are released into such polluted atmospheres, within and immediately downwind from the facility, formation of significant amounts of nitrosamines in the dark and nitramines and amides in sunlight seems possible. Clearly in these cases measurement of the actual ambient levels of amines, NO_x, HONO, nitrosamines and nitramines would be useful.

This paper is based on current research conducted at the University of California Statewide Air Pollution Research Center and the Departments of Chemistry and Biology on our campus. My colleagues include Professor W. L. Belser Jr, as well as Dr R. Atkinson, Dr K. R. Darnall, Dr R. A. Graham, Dr D. Grosjean, Dr G. B. Knudson, Dr R. Perry, Dr J. P. Schmid, Dr J. P. Smith, Dr E. C. Tuazon, Dr K. A. Van Cauwenberghe and Dr A. M. Winer, and Mr F. R. Burleson, Mr D. R. Fitz, Mr P. M. Hynds, Mr K. Pettus and Mr T. M. Mischke. Mrs Mae Minnich and Ms Yvonne Katzenstein were of great assistance in the preparation of the manuscript, including researching part of the literature. I also deeply appreciate stimulating discussions on the kinetics and photochemistry of these systems with Professor Orville Chapman and Professor B. J. Finlayson-Pitts.

I also wish to express my appreciation to the University of California and to two Federal agencies who have generously funded this work: the National Science Foundation – Research Applied to National Needs (grant no. ENV73-02904-A04, Dr R. Carrigan, project officer) and the U.S. Environmental Protection Agency (grant no. EPA804546-02) on which Dr Phil Hanst was the project officer and the designer of the F.t.–i.r. cell we employed. The contents of this paper do not necessarily reflect the views and/or policies of the N.S.F.–R.A.N.N. and the E.P.A.

Finally, I wish to thank Dr T. M. Sugden and the Royal Society for the pleasure and privilege of attending this stimulating Discussion, and the staff of the Society who were so helpful and efficient.

References (Pitts)

Ames, B. N., Durston, W. E., Yamasaki, E. & Lee, F. D. 1973 *Proc. natn. Acad. Sci. U.S.A.* **70**, 2281–2285.
Ames, B. N. & McCann, J. 1976 In *Screening tests in chemical carcinogens* (I.A.R.C. Scientific Publication No. 12), pp. 493–501. Lyon: I.A.R.C.
Ames, B. N., McCann, J. & Yamasaki, E. 1975 *Mutat. Res.* **31**, 347–364.
Asahina, J., Andrea, J., Carmel, A., Arnold, E., Bishop, Y., Joshi, S., Coffin, D. & Epstein, S. S. 1972 *Cancer Res.* **32**, 2263–2268.
Ashby, J. & Styles, J. A. 1978 *Nature, Lond.* **271**, 452–455.
Atkinson, R., Darnall, K. R., Lloyd, A. C., Winer, A. M. & Pitts, J. N. Jr 1978*a* *Adv. Photochem.* (In the press.)
Atkinson, R., Perry, R. A. & Pitts, J. N. Jr 1978*b* *J. chem. Phys.* **68**, 1850–1853.
Bamford, C. H. 1939 *J. chem. Soc.* 12–26.
Barofsky, D. F. & Baum, E. J. 1976 *J. Am. chem. Soc.* **98**, 8286–8287.
Bartsch, H. 1976 *Mutat. Res.* **38**, 177–190.
Berry, R. S. & Lehman, P. A. 1971 *A. Rev. phys. Chem.* **22**, 47–84.
Blumer, M., Blumer, W. & Reich, T. 1977 *Environ. Sci. Technol.* **11**, 1082–1084.
Bochkov, N. P., Sram, R. J., Kuleshov, N. P. & Zhurkov, V. S. 1976 *Mutat. Res.* **38**, 191–202.
Boyer, K. W. & Laitinen, H. A. 1975 *Environ. Sci. Technol.* **9**, 457–469.
Bretschneider, K. & Matz, Y. 1973 *Arch. Geschwulstforsch.* **43**, 36–42.
Calvert, J. G. & Pitts, J. N. Jr 1966 In *Photochemistry*, p. 557. New York: John Wiley & Sons.
Cautreels, W. & Van Cauwenberghe, K. A. 1976 *Atmos. Environ.* **10**, 447–457.
Chan, W. H., Nordstrom, R. J., Calvert, J. C. & Shaw, J. H. 1976 *Environ. Sci. Technol.* **10**, 674–682.
Chapman, O. L., Heckert, D. C., Reasoner, J. W. & Thackaberry, S. P. 1966 *J. Am. chem. Soc.* **88**, 5550–5554.
Commoner, B. 1976 E.P.A. Report EPA600/1-76-022. Washington, D.C.: E.P.A.
Commoner, B., Madyastha, P., Bronsdon, A. & Vithayathil, J. 1978 *J. Toxicol. environ. Hlth* **4**, 59–77.
Chrisp, C. E., Fisher, G. L. & Lammert, J. E. 1978 *Science, N.Y.* **199**, 73–75.
Daisey, J. M., Hawryluk, I., Kneip, T. J. & Mukai, F. H. 1978 Paper presented at the Conference on Carbonaceous Particles in the Atmosphere, March 20–22, Berkeley, California.

Davies, M. & Jonathan, N. 1958 *Trans. Faraday Soc.* 469–477.

Dean, B. J. 1976 *Mutat. Res.* **41**, 83–88.

Demerjian, K. L., Kerr, J. A. & Calvert, J. G. 1974 In *Advances in environmental science and technology* (ed. J. N. Pitts, Jr & R. L. Metcalf), vol. 4, pp. 1–262. New York: Wiley-Interscience.

de Serres, F. J. 1976*a* *Mutat. Res.* **38**, 355–357.

de Serres, F. J. 1976*b* *Mutat. Res.* **41**, 43–50.

Dewar, M. J. S., Mole, T., Urch, D. S. & Warford, E. W. T. 1956*a* *J. chem. Soc.* 3572–3575.

Dewar, M. J. S., Mole, T. & Warford, E. W. T. 1956*b* *J. chem. Soc.* 3581–3586.

Dong, M., Locke, D. & Hoffmann, D. 1977 *Environ. Sci. Technol.* **11**, 612–618.

Doyle, G. J., Bekowies, P. J., Winer, A. M. & Pitts, J. N. Jr 1977 *Environ. Sci. Technol.* **11**, 45–51.

Druckrey, H., Preussman, R., Schmahl, D. & Muller, M. 1961 *Naturwissenschaften* **48**, 134.

Epstein, S. S. 1974 *Cancer Res.* **34**, 2425–2435.

Epstein, S. S., Joshi, S., Andrea, J., Mantel, N., Sawicki, E., Stanley, T. & Tabor, E. C. 1966 *Nature, Lond.* **212**, 1305–1307.

Epstein, S. S., Mantel, N. & Stanley, T. W. 1968 *Environ. Sci. Technol.* **2**, 132–138.

Fahmy, O. G. & Fahmy, M. J. 1976 *Cancer Res.* **36**, 4504–4512.

Falk, H. L., Markul, I. & Kotin, P. 1956 *A.M.A. Archs ind. Hlth* **13**, 13–17.

Fatiadi, A. J. 1967 *Environ. Sci. Technol.* **1**, 10–12.

Fine, D. H. 1978 In *Advances in environmental science and technology* (ed. J. N. Pitts, Jr & R. L. Metcalf), vol. 9. New York: Wiley-Interscience. (In the press.)

Fine, D. H., Rounbehler, D. P., Belcher, N. M. & Epstein, S. S. 1976 *Science, N.Y.* **192**, 1328–1330.

Fine, D. H., Rounbehler, D. P., Rounbehler, A., Silvergleid, A., Sawicki, E., Krost, K. & DeMarrais, G. A. 1977 *Environ. Sci. Technol.* **11**, 581–584.

Finlayson, B. J. & Pitts, J. N. Jr 1976 *Science, N.Y.* **192**, 111–119.

Finlayson-Pitts, B. J. & Pitts, J. N. Jr 1977 In *Advances in environmental science and technology* (ed. J. N. Pitts, Jr & R. L. Metcalf), vol. 7, pp. 75–163. New York: Wiley-Interscience.

Fishbein, L. 1976 In *Chemical mutagens* (ed. A. Hollaender), vol. 4, pp. 219–319. New York: Plenum Press.

Freeman, A. E., Price, P. J., Bryan, R. J., Gordon, R. J., Gilden, R. V., Kelloff, G. J. & Huebner, R. J. 1971 *Proc. natn. Acad. Sci. U.S.A.* **68**, 445–449.

Friedlander, S. K. & Miguel, A. 1978 *Atmos. Environ.* (In the press.)

Goodall, C. M. & Kennedy, T. H. 1976 *Cancer Lett.* **1**, 295–298.

Gordon, R. J. & Bryan, R. J. 1973 *Environ. Sci. Technol.* **7**, 1050–1053.

Gordon, R. J., Bryan, R. J., Rhim, J. S., Demoise, C., Wolford, R. G., Freeman, A. E. & Huebner, R. J. 1973 *Int. J. Cancer* **12**, 223–227.

Goulden, F. & Tipler, M. M. 1949 *Br. J. Cancer* **3**, 157–160.

Grossman, B. 1976 Ph.D. Thesis, University of Antwerp.

Grimmer, G. 1977 *International Agency for Research on Cancer* (*I.A.R.C.*) *Sci. Pub.* no. 16, pp. 29–39. Lyon, France.

Guether, B. 1864 *Arch. Pharmac.* **123**, 200.

Hanst, P. L., Spence, J. W. & Miller, S. M. 1977 *Environ. Sci. Technol.* **11**, 403–405.

Heicklen, J. 1976 *Proc. Conference on Ozone/Oxidants – Interactions with the Total Environment*, Publ. SP-7. Pittsburgh, Pennsylvania: Air Pollution Control Association.

Heidelberger, C. 1975 *A. Rev. Biochem.* **44**, 79–121.

Hoffmann, D. & Wynder, E. L. 1977 In *Air pollution* (ed. A. C. Stern), third edn, vol. 2, pp. 361–455. New York: Academic Press.

Hueper, W. C., Kotin, P., Tabor, E. C., Payne, W. W., Falk, H. L. & Sawicki, E. 1962 *Archs Path.* **74**, 89–116.

Huisingh, J., Bradow, R., Jungers, R., Claxton, L., Zweidinger, R., Tejada, S., Bumgarner, J., Duffield, F., Waters, M., Simmon, V., Hare, C., Rodriguez, C. & Snow, L. 1978 Paper presented at the Symposium on Application of Short-Term Bioassays in the Fractionation and Analysis of Complex Environmental Mixtures, February 21–23, Williamsburg, Virginia.

Inomata, M. & Nagata, C. 1972 *Gann* **63**, 119–130.

Jaeger, J. 1971 *Fresenius' Z. Chem.* **225**, 281–284.

Jäger, J. 1978 *J. Chromat.* **152**, 575–578.

Jackson, B. & Dessau, F. I. 1961 *Lab. Invest.* **10**, 909–923.

Jerina, D. M., Lehr, R. E., Yagi, H., Hernandez, O., Dansette, P. M., Wislocki, P. G., Wood, A. W., Chang, R. L., Levin, W. & Conney, A. H. 1976*a* In *In vitro metabolic activation in mutagenesis testing* (ed. F. J. de Serres, J. R. Fouts, J. R. Bend & R. M. Philpot), pp. 159–177. Amsterdam: Elsevier.

Jerina, D. M., Yagi, H., Hernandez, O., Dansette, P. M., Wood, A. W., Levin, W., Chang, R. L., Wislocki, P. G. & Conney, A. H. 1976*b* In *Polynuclear aromatic hydrocarbons: chemistry, metabolism and carcinogenesis* (ed. R. I. Freudenthal & P. W. Jones). pp. 91–113. New York: Raven Press.

Kaiser, E. W. & Wu, C. H. 1977 *J. phys. Chem.* **81**, 1701–1706.

Kapitulnik, J., Levin, W., Yagi, H., Jerina, D. & Conney, A. 1976 *Cancer Res.* **36**, 3625–3628.

Kertesz-Saringer, M., Meszaros, E. & Varkonyi, T. 1971 *Atmos. Environ.* **5**, 429–431.

Kotin, P., Falk, H. L. & McCammon, C. J. 1958 *Cancer* **11**, 473–481.

Kotin, P., Falk, H. L., Mader, P. & Thomas, M. 1954 *Archs ind. Hyg.* **9**, 153–163.
Kotin, P., Falk, H. L. & Thomas, M. 1956 *Cancer* **9**, 905–909.
Lane, D. 1976 Ph.D. thesis, York University, Toronto.
Lao, R. C., Thomas, R. S., Oja, H. & Dubois, L. 1973 *Analyt. Chem.* **45**, 908–915.
Larson, R. A., Hunt, L. L. & Blankenship, D. W. 1977 *Environ. Sci. Technol.* **11**, 492–496.
Leiter, J. & Shear, M. J. 1943 *J. natn. Cancer Inst.* **3**, 455–477.
Leiter, J., Shimkin, M. B. & Shear, M. J. 1942 *J. natn. Cancer Inst.* **3**, 155–165.
Lesclaux, R., Khê, P. V., Desauzier, P. & Soulignac, J. C. 1975 *Chem. Phys. Lett.* **35**, 493–497.
Levine, S. Z. & Calvert, J. G. 1977 *Chem. Phys. Lett.* **46**, 81–84.
Litman, R. 1978 *Environ. Sci. Technol.* **12**, 599.
Lu, A., Levin, W., Vore, M., Conney, A. H., Thakker, D. R., Holder, G. & Jerina, D. M. 1976 In *Polynuclear aromatic hydrocarbons: chemistry, metabolism and carcinogenesis* (ed. R. I. Freudenthal & P. W. Jones), pp. 115–126. New York: Raven Press.
McCann, J. 1976 *ChemTech*, pp. 682–687.
McCann, J., Choi, E., Yamasaki, E. & Ames, B. N. 1975 *Proc. natn. Acad. Sci. U.S.A.* **72**, 5135–5139.
Magee, P. N. & Barnes, J. 1956 *Br. J. Cancer* **10**, 114–122.
Masuda, Y. & Kuratsune, M. 1966 *Air Water Pollut.* **10**, 805–811.
Maugh, T. H., II. 1976 *Science, N.Y.* **193**, 871–873.
Mohr, U., Reznik-Schuller, H., Reznik, G., Grimmer, G. & Misfeld, J. 1976 *Zbl. Bakt. Hyg.* (*I Abt. Orig. B*) **163**, 425–432.
National Academy of Sciences 1972 *Particulate organic matter*. Washington, D.C.: National Academy of Sciences.
National Academy of Sciences 1977 *Ozone and other photochemical oxidants*. Washington, D.C.: National Academy of Sciences.
Natusch, D. F. & Wallace, J. R. 1974 *Science, N.Y.* **186**, 695–699.
Passey, R. D. 1922 *Br. med. J.* ii, 1112–1113.
Pellizzari, E. D. 1977 National Technical Information Report, no. PB-269 582.
Perry, R. A., Atkinson, R. & Pitts, J. N. Jr 1977 *J. phys. Chem.* **81**, 296–304.
Pierce, R. C. & Katz, M. 1975 *Environ. Sci. Technol.* **9**, 347–353.
Pierce, R. C. & Katz, M. 1976 *Environ. Sci. Technol.* **10**, 45–51.
Pitts, J. N. Jr 1975 Second Annual Report, National Science Foundation – Research Applied to National Needs Grant No. AEN73-02904-A02, p. V-8.
Pitts, J. N. Jr, Khan, A. U., Smith, E. B. & Wayne, R. P. 1969 *Environ. Sci. Technol.* **3**, 241–247.
Pitts, J. N. Jr, Grosjean, D., Mischke, T. M., Simmon, V. F. & Poole, D. 1977*a* *Toxicol. Lett.* **1**, 65–70.
Pitts, J. N. Jr, Smith, J. P., Fitz, D. R. & Grosjean, D. 1977*b* *Science, N.Y.* **197**, 225–257.
Pitts, J. N. Jr, Belser, W. L. Jr, Van Cauwenberghe, K. A., Grosjean, D., Schmid, J. P., Fitz, D. R., Knudson, G. B. & Hynds, P. M. 1978*a* Paper presented at the Symposium on Application of Short-Term Bioassays in the Fractionation and Analysis of Complex Environmental Mixtures, February 21–23, Williamsburg, Virginia.
Pitts, J. N. Jr, Grosjean, D., Van Cauwenberghe, K. A. & Belser, W. L. Jr 1978*b* Division of Environmental Chemistry, Symposium on Chemical and Biological Implications of Nitrogenous Air Pollutants, 175th National ACS Meeting, Anaheim, California.
Pitts, J. N. Jr, Van Cauwenberghe, K. A., Grosjean, D., Schmid, J. P., Fitz, D., Belser, W. L. Jr, Knudson, G. B. & Hynds, P. M. 1978*c* *Science, N.Y.* (In the press.)
Pitts, J. N. Jr, Grosjean, D., Van Cauwenberghe, K. A., Schmid, J. P. & Fitz, D. 1978*d* *Environ. Sci. Technol.* **12**, 946–953.
Pitts, J. N. Jr, Grosjean, D., Mischke, T. M., Simmon, V. F. & Poole, D. 1978*e* In *Biological effects of environmental pollutants* (ed. S. D. Lee). Ann Arbor, Michigan: Ann Arbor Science Pub. (In the press.)
Pott, P. 1775 In *Chirurgical observations*, pp. 63–68. London: Hawes, Clarke & Collins.
Rigdon, R. H. & Neal, J. 1971 *Texas reports on biology and medicine* **29**, 110–123.
Sawicki, E. 1967 *Archs environ. Hlth* **14**, 46–53.
Sawicki, E., McPherson, S. P., Stanley, T. W., Meeker, J. & Elbert, W. C. 1965 *Int. J. Air Water Pollut.* **9**, 515–524.
Searle, C. E. 1976 *Chemical carcinogens*. American Society Monograph, no. 173. Washington, D. C.: American Chemical Society.
Smith, P. A. S. & Loeppky, R. N. 1967 *J. Am. chem. Soc.* **89**, 1147–1157.
Sobels, F. H. 1977 *Mutat. Res.* **46**, 245–260.
Spindt, R. S. 1974 First annual report on polynuclear aromatic content of heavy duty diesel exhaust gases, Gulf Research and Development Co.
Sugimura, T., Yahagi, T., Nagao, M., Takenchi, M., Kawachi, T., Hara, K., Gamasaki, L., Matsushima, T., Hashimoto, Y. & Okada, M. 1976 In *Screening tests in chemical carcinogens* (ed. R. Montesano, H. Hartsch & L. Tomatis), pp. 81–101. Lyon: I.A.R.C. Scientific Publication No. 12.
Suryanarayanan, K. & Bulusu Suryanarayanana, A. 1972 *J. phys. Chem.* **76**, 496–500.
Talcott, R. & Wei, E. 1977 *J. natn. Cancer Inst.* **58**, 449–451.

Tebbens, B. D., Mukai, M. & Thomas, J. F. 1971 *J. Am. ind. Hyg. Ass.* **32**, 365–372.
Tebbens, B. D., Thomas, J. F. & Mukai, M. 1966 *J. Am. ind. Hyg. Ass.* **27**, 415–422.
Teranishi, K., Hamada, K. & Watanabe, H. 1978 *Mutat. Res.* **56**, 273–280.
Tokiwa, H., Tokeyoshi, H., Morita, K., Takahashi, K., Soruta, N. & Ohnishi, Y. 1976 *Mutat. Res.* **38**, 351–359.
Tuazon, E. C., Winer, A. M., Graham, R. A. & Pitts, J. N. Jr 1977 Paper presented at 4th Joint Conference on Sensing of Environmental Pollutants, November, 1977, New Orleans, Louisiana. (In the press.)
Tuazon, E. C., Winer, A. M., Graham, R. A., Schmid, J. P. & Pitts, J. N. Jr 1978*a* *Environ. Sci. Technol.* **12**, 954–958.
Tuazon, E. C., Graham, R. A., Winer, A. M., Easton, R. R., Pitts, J. N. Jr & Hanst, P. L. 1978*b* *Atmos. Environ.* **12**, 865–875.
U.S. Environmental Protection Agency 1976 *Scientific and technical assessment report on nitrosamines.* Report No. EPA-600/6-77-001, Washington, D.C.
U.S. Environmental Protection Agency 1977 *Reconnaissance of environmental levels of nitrosamines in the Central United States.* EPA Document 330/1-77-001, EPA National Enforcement Investigations Center, Denver, Colorado.
Waller, R. E. 1952 *Br. J. Cancer* **6**, 8–21.
Walters, C. L. 1977 *Chem. Br.* **13**, 140–145.
Wang, Y.-Y., Sawyer, R. F. & Wei, E. T. 1978 Paper presented at the Symposium on Application of Short-Term Bioassays in the Fractionation and Analysis of Complex Environmental Mixtures, February 21–23, Williamsburg, Virginia.
Weisburger, J. H., Yamamoto, R. S., Glass, R. M. & Frankel, H. H. 1969 *Toxicol. appl. Pharmac.* **14**, 163–175.
Winer, J., Hynds, P. M., Shaffer, S., Belser, W. L. Jr & Pitts, J. N. Jr 1978 Submitted to *Mutat. Res.*
Wislocki, P., Wood, A., Chang, R., Levin, W., Yagi, H., Hernandez, O., Dansette, P., Jerina, D. & Conney, A. 1976 *Cancer Res.* **36**, 3350–3357.
Wynder, E. L. & Hoffmann, D. 1965 *J. Air Pollut. Control Ass.* **15**, 155–158.

Phil. Trans. R. Soc. Lond. A. **290**, 577–589 (1979)
Printed in Great Britain

The dispersion of lead from motor exhausts

BY A. C. CHAMBERLAIN, M. J. HEARD, P. LITTLE
AND R. D. WIFFEN
Environmental and Medical Sciences Division, Atomic Energy Research Establishment, Harwell, Oxfordshire OX11 0RA, U.K.

[Plate 1]

Lead is added to petrol as organic tetraethyl or tetramethyl lead but is emitted as inorganic oxides, sulphides, halides and carbonates. From vehicles cruising at high speed on motorways, the lead is emitted as very small (*ca.* 0.02 μm) discrete particles. In city streets, where there is also a background aerosol, coagulated chain-aggregate particles are found.

The concentration of lead in air downwind of a road can be calculated theoretically, and measured concentrations agree fairly well. Within a few metres of the carriageways of motorways carrying heavy traffic, concentrations as high as 10 $\mu g/m^3$ are found, but the concentration falls rapidly with distance owing mainly to the upward diffusion of the plume. Measurements of the rate of deposition to surfaces near the M 4 motorway show that an appreciable, but not large, fraction of the lead is deposited within 100 m of the motorway.

Near towns, the contribution from a given highway merges rapidly into the background due to emissions from other roads. The dispersion is countrywide and probably some cross-frontier transfer occurs.

1. EMISSION FROM VEHICLES

Lead is added to petrol as the organic liquids tetraethyl lead and tetramethyl lead. About 75 % of the lead is emitted from the exhaust (references in Huntzicker, Friedlander & Davidson 1975). The lead is in a variety of chemical forms (halides, oxides, carbonates and sulphates). Measurements in London streets (Harrison, Perry & Slater 1974) show that organic lead vapour accounts for 0.1–4.0 % of total lead. Near filling stations the percentage is higher. Most of the organic lead is thought to come from evaporative losses.

The mean particle size (mass median equivalent diameter, m.m.e.d.) of lead in aerosols in urban air has been reported to be 0.2–0.4 μm (Cholak, Schafer & Yeager 1968). These results are derived from measurements with impactors, but as more than half the lead is usually found on the backing filter paper it is necessary to extrapolate the readings to obtain the m.m.e.d. Similar measurements currently being made on the urban aerosol at Chelsea (London) extrapolate to an m.m.e.d. of 0.27 μm but the assumed log-normal distribution has a rather large geometric deviation ($\sigma_g = 7.3$) and it is doubtful if this type of interpretation is meaningful. The interpretation of impactor data is even more doubtful when applied to the motorway aerosol, resulting in an m.m.e.d. below 0.1 μm, but values of σ_g in excess of 50 suggest a complex distribution which is not resolved by the impactor techniques.

Figure 1, plate 1, shows electron micrographs of aerosol collected by thermal precipitator alongside the M 4 near Heston, Middlesex. Three different magnifications are shown. Samples taken upwind of the motorway showed very much fewer particles.

There are three principal types of particle:

(*a*) round, electron dense particles in the size range 0.01–0.1 μm, some single and some multiple, which decompose in the electron beam in a manner which appears to be characteristic of primary exhaust particles;

(*b*) chain aggregates of less electron dense particles, 0.3–1.0 μm, mostly diesel smoke;

(*c*) larger, often scaly, particles, greater than 1 μm, probably road dust and adventitious material.

Similar particles can be seen in the electron micrographs of Lawther, Commins, Ellison & Biles (1973).

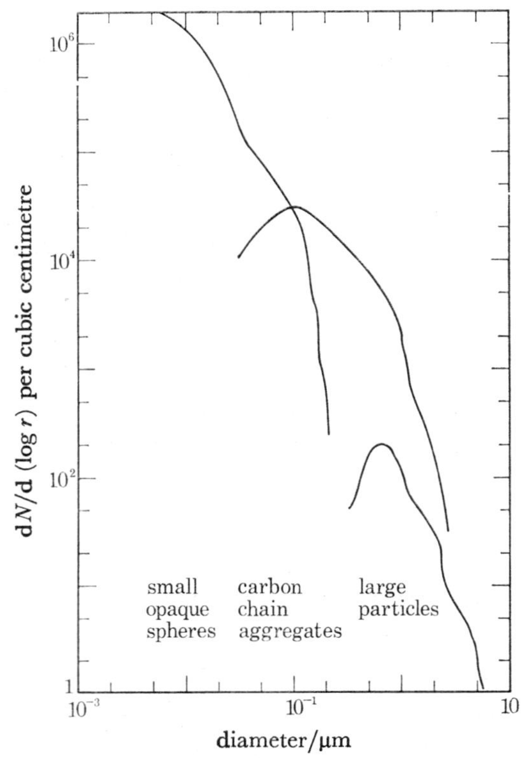

FIGURE 2. Vehicle aerosols: size distributions of three classes of particle; M 4, Heston.

Figure 2 shows the particle size spectrum obtained from the electron micrographs. To confirm that the smallest particles contained most of the lead, a diffusion battery (Megaw & Wiffen 1963) was used to estimate the Brownian diffusivity of the lead-bearing particles. In a number of comparisons, the percentage penetration of lead aerosol through the battery gave diffusion mass median diameters of 0.03–0.05 μm corresponding to electron micrograph volume median diameters from 0.04–0.08 μm. The Brownian diffusivity of a particle, unlike its impaction efficiency, does not depend on its density (Fuchs 1964, p. 184). Also, the mass of lead which would be expected to be associated with the small particles agreed in order of magnitude with the measured air concentration. Measurements near a freeway in the U.S.A. with an electrical analyser (Whitby *et al.* 1975) had previously shown that most of the volume distribution of primary exhaust aerosols lies in the nucleus mode (0.01–0.1 μm) with a peak at 0.03 μm. Wind tunnel work with a small engine indicates that the primary exhaust particle is approximately 0.015 μm diameter. The eventual size distribution depends on the rate of

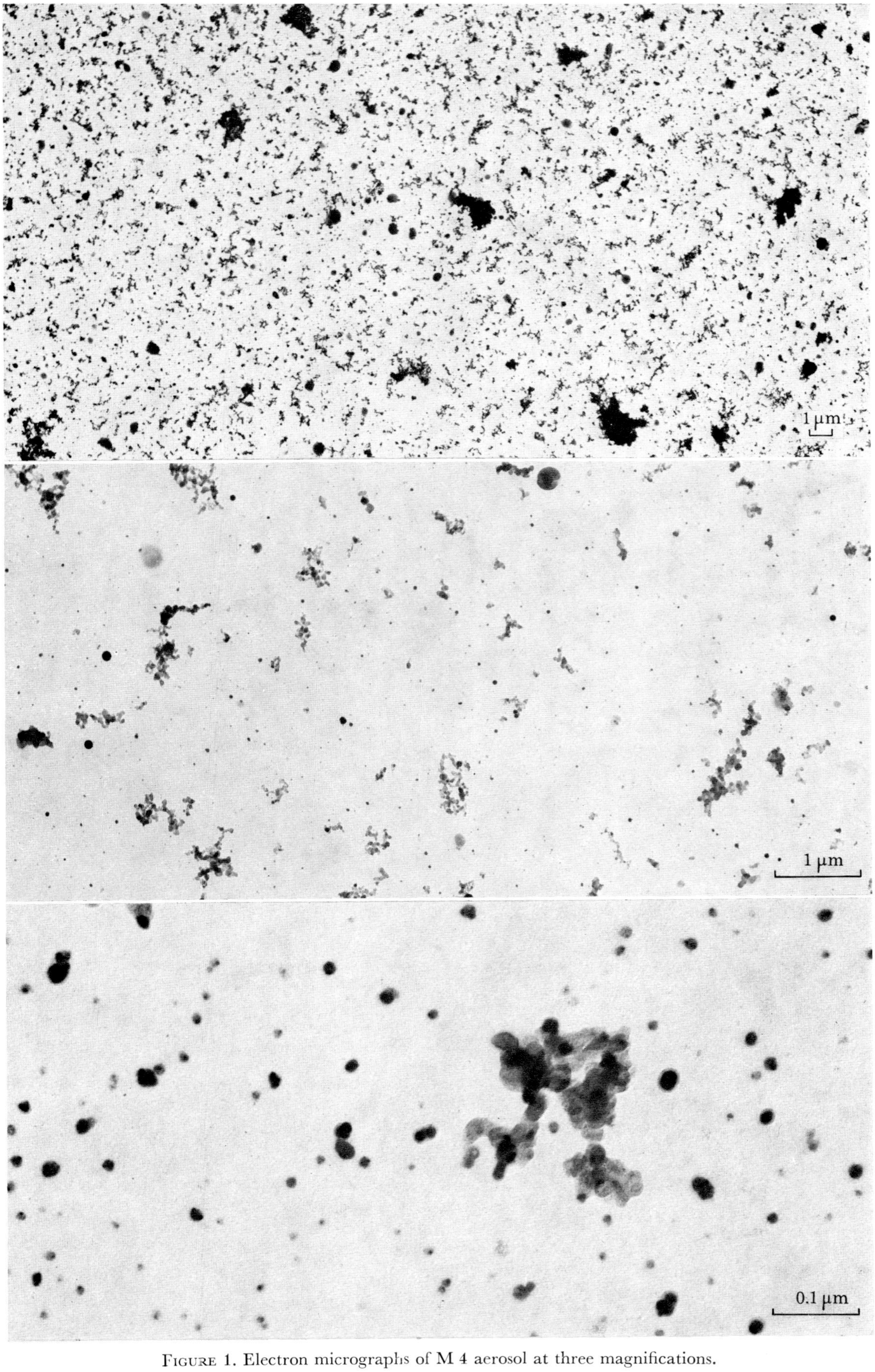

FIGURE 1. Electron micrographs of M 4 aerosol at three magnifications.

(*Facing p.* 578)

dilution of the exhaust, and the number and size of other aerosol particles with which the lead particles may coagulate. Measurements near the M 40, where the motorway ascends the Chiltern escarpment on a 1 in 20 grade, showed little difference in the size spectrum compared with the M 4, but measurements at the Hogarth roundabout at the end of the M 4 in Chiswick showed a shift in the distribution with more particles in the 0.1–0.5 and 0.5–5 μm ranges. This is attributed mainly to coagulation with the general urban aerosol, but differences in driving modes may also have an effect.

As the particle size is small, the dispersion of exhaust lead can be considered in terms of the well established theory of diffusion of gases and smokes. Dispersion is determined by wind and turbulence, since the sedimentation velocity of the particles is smaller compared with turbulent air velocities. Deposition on roadside verges, which is considered later, is secondary to dispersion in determining the air concentration downwind of highways. The emission of lead from cars on a roadway can be calculated from the flow of traffic, lead content of petrol, petrol consumption and percentage emission of lead. Where traffic flow is given this will be petrol-driven vehicles per hour unless otherwise stated. Assuming 1000 vehicles per hour, 0.49 g Pb l^{-1} in petrol, fuel consumption 0.24 l km^{-1} (30 miles per gallon) and 75 % emission, the source strength q is 10 μg s^{-1} per metre length of roadway. This result can also be derived from measurements at Warren Spring Laboratory of the emissions from a car driven in traffic (Hogbin & Bevan 1976), if the average velocity is taken as 60 miles per hour (*ca.* 97 km h^{-1}).

2. Lead in air near roads

Cars travelling on a lane of a motorway create a line source of lead in air, and if the wind is perpendicular to the motorway the concentration χ near ground level downwind can be calculated using established formulae (Pasquill 1974). If cars travel equally spaced on all lanes,

$$\chi_{(x)} = (2/\pi)^{\frac{1}{2}} \frac{q}{uw} \int_{x}^{x+w} \frac{dx}{\sigma_z(x)}, \tag{1}$$

where q is the source strength of Pb (in micrograms per metre per second), u the wind speed in metres per second, w the width in metres of motorway between edge beams, taken as 25 m (edge beam separates carriageway from hard shoulder), x the distance in metres of point of measurement from edge beam, and $\sigma_z(x)$ the standard deviation in metres of vertical spread of plume ($= ax^s$, where a, s are parameters depending on atmospheric stability; Pasquill, 1974, tab. 6, IX).

In figure 3, χ is plotted for stability categories B (unstable), D (neutral) and E (stable). A wind speed of 2 m s^{-1} has been assumed. This is close to the harmonic mean wind speed at height 1 m at inland stations in southern England. Figure 3 shows that atmospheric stability in the range of categories B–E accounts for a 1:3 range in calculated χ. The most extreme stable categories F and G are thought unlikely to apply since the vehicle motion and heat of exhausts creates initial dispersion. Allowing also for variations in wind speed, a variation by a factor of ten at least can be expected. There are also further variables difficult to allow for:

(*a*) Roughness of terrain: calculations of figure 3 assume roughness length $z_0 = 0.1$ m. In towns z_0 may be about 1 m, which will increase dispersion and reduce χ by about 50 %, but in very enclosed urban situations, with a canyon effect, χ may be increased.

(*b*) Angle of wind to motorway: Calder (1973) has shown that the angle, θ, makes little difference unless it exceeds 60° to the perpendicular, i.e. wind nearly parallel to road. Then χ is increased and at $\theta = 75°$ the increase is about 50% relative to $\theta = 0°$.

(*c*) Elevation of motorway: if the roadway is elevated by say 10 m, then mixing while the plume descends to ground level, including the effect of eddies caused by the roadway structure, will result in a minimum σ_z of about 5 m. This means that χ at ground level will be a maximum when x is about 100 m, and will not show the rapid increase for small x of figure 3. Conversely, in a cutting, particularly under stable conditions, dispersion will be poor.

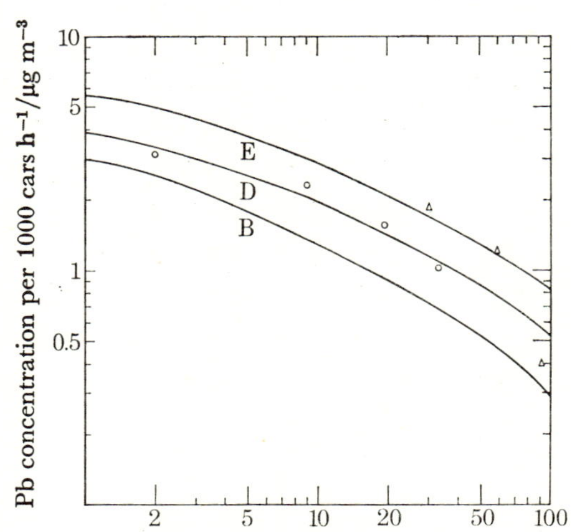

FIGURE 3. Lead in air plotted against distance from motorway. Lines are labelled with Pasquill stability categories. Measurements: ○, Little & Wiffen (1977); △, Bevan *et al.* (1974).

The points in figure 3 are derived from measurements by Bevan, Colwill & Hogbin (1974) and by Little & Wiffen (1978) of lead in air alongside the M 4 motorway at Heston, Middlesex. This is the busiest length of motorway in the U.K., with about 4000 vehicles per hour during much of the day. Bevan *et al.* measured Pb in the central reservation of the M 4 continuously on weekdays from 30 August to 27 October 1972, and showed that the concentration correlated closely with traffic volume. They also ran samplers for several days at each of four distances (30, 60, 90 and 120 m from the edge beam) and compared the results with simultaneous samplers from the central reservation. From the ratios so derived, the mean χ (central) (15.1 µg/m³), and the mean traffic flow (3800 vehicles per hour), the points in figure 3 are derived. Little & Wiffen sampled at 2, 9, 19 and 34 m from the edge beam for about 10 h on each of 3 days, and also did traffic counts. The values of χ normalized to 1000 vehicles per hour so derived are also shown in figure 3. The stability categories for these runs were B (twice) and C/D. For the runs performed by Bevan *et al.*, no estimate of stability is available. At the time of their work the maximum lead content of petrol was 0.84 g/l, whereas at the time of Little & Wiffen's work it was 0.50 g/l, and if allowance is made for this difference the two sets of results agree well. On 1 January 1978 the maximum was further reduced to 0.45 g/l. The agreement of the experimental points in figure 3 with theory, both in absolute magnitude and in variation of χ with x, give confidence that dispersion of exhaust lead is similar to that of other smokes.

TABLE 1. MEASUREMENTS OF LEAD IN AIR NEAR MOTORWAYS AND OTHER STREETS

location	site type†	sampling periods	petrol vehicles per hour	lead in air/μg m^{-3} measured	lead in air/μg m^{-3} background (estimated)	χ μg m^{-3} per 10^3 vehicles h^{-1}	reference
M 4, central reservation	O	24 h	3800	15.1	0.4	3.9	Bevan *et al.* (1974)
M 4, 2 m from edge beam	O	daytime	3600	8.9	0.4	2.4	Little & Wiffen (1978)
M 40, 2 m from edge beam level	O	daytime	970	3.3	0.2	3.2	
1/20 grade in cutting	C	daytime	950	10.3	0.2	10.6	
1/20 grade on embankment	O	daytime	990	6.4	0.2	6.2	
London							
Fleet Street, central constricted	C	daytime	1200	3.2	0.8	2.0	Waller *et al.* (1965)
Exhibition Rd, S.W.7 kerbside	O	daytime	1300	3.2	0.5	2.0	Harrison *et al.* (1974)
Seymour Place, W.1	C	daytime	424	4.0	1.0	7.1	
Talgarth Rd, W.14	O	daytime	3880	8.8	0.5	2.2	
Upper Berkeley St, W.1	C	daytime	420	4.3	1.0	7.9	
Lancaster, A 6	C	24 h	830	3.5	0.5	3.6	Harrison (1977)

† O, open site; C, constricted site

The reduction of χ with increasing x is independent of loss by deposition, which is considered below. In an urban area, where the background lead in air from the whole network of roads is likely to be 0.5–1.0 µg m^{-3}, the effect of one particular highway becomes difficult to distinguish beyond about 100 m from the carriageway. In table 1, other roadside measurements of lead in air are compared with those from the M 4. The M 40 samples were taken where the motorway ascends the Chilton escarpment on a 1 in 20 grade partly on embankment and partly in cutting. The χ values normalized to 1000 vehicles per hour show the effects of increased petrol consumption uphill and of the confinement of the cutting. The particle size spectrum was little different from that near the M 4. Other measurements of lead in air are also compared in table 1. At open sites with freely moving traffic, the normalized χ is about 2 µg m^{-3}, but in narrow city streets, with slow moving traffic, it is higher.

3. Lead in air from areal sources

The mean concentration of lead in air over a city or district will now be considered in terms of the mean source strength per unit area. The concentration C near ground at the centre of a circular area radius x_2 in which the source strength is Q µg m^{-2} s^{-1} is

$$C = \int_{x_1}^{x_2} \left(\frac{2}{\pi}\right)^{\frac{1}{2}} \frac{Q}{u} \frac{dx}{\sigma_z(x)};$$

$$C/Q = \left(\frac{2}{\pi}\right)^{\frac{1}{2}} \frac{(x_2^{1-s} - x_1^{1-s})}{ua(1-s)}, \tag{2}$$

where appropriate values of u, a and s are to be taken according to the meteorological conditions.

The lower limit of integration x_1, is taken as 50 m, as the measurements to be compared, unlike those of table 1, are made typically at sites not immediately adjacent to roads.

Table 2. Areal emissions of lead and concentrations in air

region	area km^2	equiv. diam. km	emission of Pb (t/a)	emission of Pb $Q/(10^{-2}$ µg m^{-2} s^{-1})	mean Pb in air, C µg m^{-3}	C/Q s m^{-1}
Birmingham C.B.	2.1×10^2	16	125	1.9	0.75	39
Inner London	2.3×10^2	17	270	3.8	0.8	22
Los Angeles County	4.4×10^3	75	6400	4.6	2.4	52
U.K.	1.7×10^5	465	9000	1.7	0.14	82

References

Emissions: U.K., Stubbs (1972); Inner London, Ball & Hume (1977); Birmingham, M. L. Williams (personal communication); Los Angeles, Huntzicker *et al.* (1975).

Concentrations: U.K., Cawse (1974, 1975), Cambray *et al.* (1975); London, Ball & Hume (1977), Lawther *et al.* (1973); Birmingham, Archer & Barratt (1976); Los Angeles, Huntzicker *et al.* (1975).

In table 2, literature values of C and Q for Inner London, the U.K. and Los Angeles are given. A statistical analysis of Pasquill stability categories according to time of day, from observations over 10 years at London, Birmingham and Bristol, was supplied by the Meteorological Office and used to calculate C/Q on an annual mean as a function of the diameter of the area considered. The diurnal frequency of stability categories was weighted by the diurnal variation of traffic flow. This is important because the most stable categories E and F, for which dispersion is poor and C/Q large, occur mainly at night when traffic is lighter than by day. Table 3 shows that the weighted frequencies of categories E and F are about half the unweighted.

TABLE 3. FREQUENCY OF STABILITY CATEGORIES

category	A	B	C	D	E	F
wind speed (m s^{-1}) at height 10 m	1.0	1.4	3.5	5.6	2.8	1.4
% frequency						
as observed	0.8	6.0	15.4	61.0	7.2	9.7
as weighted by diurnal variations of traffic density	1.3	8.1	22.1	60.4	3.9	4.3

Figure 4 shows good agreement between calculated and measured values of C/Q. No allowance has been made for contributions to C from emissions outside the areas considered nor for deposition.

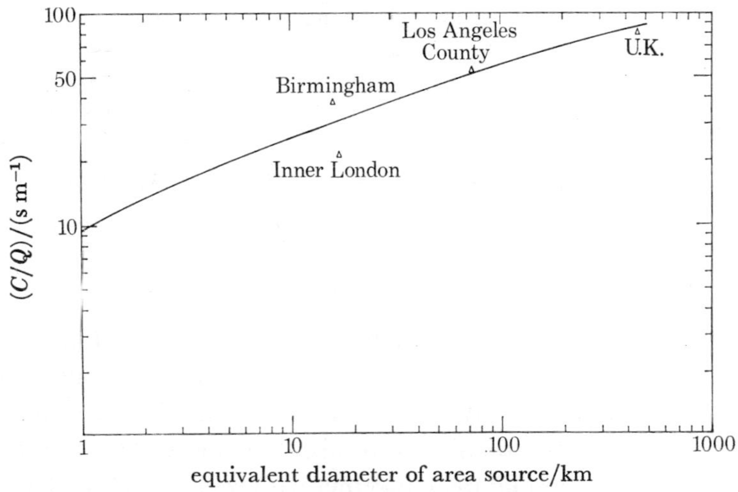

FIGURE 4. Lead in air from area source. The line is calculated from equation (2) with the use of table 3.

Owing to the shorter daylight, categories E and F are more frequent in winter than in summer. Calculations were done for the seasons separately, and gave the ratio of C/Q in winter:summer to be about 1.2:1. Measurements of C have given winter:summer ratios of 1.6 in rural areas (Cawse 1974) and about 2 in cities (Ball & Hume 1977; Day, Evans & Robson 1977). It is not known why the observed seasonal variation is larger than calculated. The seasonal variation of C provides a possible method of estimating the contribution of lead in air to lead in blood for the population, by seeing if the latter shows seasonal variations.

For comparison with the results for lead, table 4 shows estimates of C/Q for the mean rural concentrations of SO_2 and NO_x in the U.K. C/Q for SO_2 is lower than C/Q for Pb, probably because SO_2 is deposited more rapidly on the ground surface, and also converted to SO_4^{2-} by photochemical and other routes. C/Q values of 34 and 84 s m^{-1} for SO_2 in Los Angeles and New York respectively can be deduced from a table of emissions and concentrations given by Hidy, Mueller & Tang (1977).

Figure 4 can be used to estimate the contributions to lead in an urban area from cars on a motorway through it. The total source strength of 1 km of motorway with 1000 vehicles per hour is 10000 μg s^{-1}, and this contributes 0.01 μg m^{-2} s^{-1} to Q for the area of 1 km^2 comprising a strip 500 m wide on either side of the motorway. From figure 4 this will contribute about 0.1 μg m^{-3} to the mean lead in air over this area. This is unlikely to be detected statistically against an urban background typically in the range 0.5–1.0 μg m^{-3}. For example, Archer & Barratt (1976) estimate that 9.5 kg of lead are discharged weekly from cars within a radius of 0.25 miles of the M 6 interchange at Gravelly Hill in Birmingham. This gives $Q = 0.03$ μg m^{-2} s^{-1} and $C \approx 0.3$ μg m^{-3} as the contribution of the local traffic. Measurements (Archer & Barratt 1976; C.U.E.P. 1975) in 1973–4 gave $C = 1.5$ μg m^{-3} in the centre of the Salford Circus roundabout near the centre of the interchange and $C = 1.0$, 0.7 and 0.9 μg m^{-3} near houses on the periphery of the interchange. The mean C in Birmingham generally was 0.75 μg m^{-3} (Archer & Barratt 1976). Lead levels in roadside dust near the Gravelly Hill interchange did not differ significantly from those in the city generally.

TABLE 4. COMPARISON OF NORMALIZED CONCENTRATIONS OF POLLUTANTS

pollutant	emissions in U.K. $t\ a^{-1}$	emissions in U.K. μg m^{-2} s^{-1}	mean conc. μg m^{-3}	C/Q s m^{-1}	reference
Pb	9000	1.7×10^{-3}	0.14	80	
smoke	2.3×10^{6}	0.43	57	133	Meetham (1950)
SO_2	6×10^{6}	1.1	36	33	Garland (1977)
NO_x (measured at NO)	1.4×10^{6}	0.26	10.6†	41	Derwent & Stewart (1973)

† Mean concentration at Harwell, 1973–5 (Cox, Derwent & Sandalls 1976). Levels for SO_2 and Pb at Harwell are close to U.K. non-urban averages.

4. DEPOSITION OF LEAD

Little & Wiffen (1978) measured the deposition of lead beside the M 4 motorway by exposing shallow trays of grass, 10 cm high, placed on the central reservation and at several distances upwind and downwind of the motorway. An exposure period of 3 h was enough to measure deposition. Wind speed and traffic density were also measured. The results, normalized to 1000 vehicles per hour, are shown in figure 5, together with results of Huntzicker *et al.* (1975 and private communication) obtained near a freeway in Los Angeles. The ratio of the rate of deposition (μg cm^{-2} s^{-1}) to the air concentration (μg cm^{-3}) is the velocity of deposition (v_g). This was found by Little & Wiffen to decrease from 1.0 cm s^{-1} at 2 m from the edge beam to 0.24 cm s^{-1} at 34 m. The higher v_g very near to the motorway was probably due mainly to the turbulence caused by nearby vehicles passing at speed. Little & Wiffen estimated from their deposition measurements that about 9 % of the lead emitted from vehicles on the motorway was deposited within 100 m. Huntzicker *et al.* found 6 % deposited within 150 m. There is a possibility of additional deposition of large particles derived from the surface of exhaust systems, on the roadway surface itself, particularly where cars accelerate rapidly. Cawse (1974, 1975), and Cambray, Jefferies & Topping (1975) measured both the air concentrations and the deposition of lead at a number of rural and coastal sites in the U.K. The total (wet and dry) deposition collecting in funnels was measured and also the dry deposition to horizontal filter pads, sheltered from rain. These pads collected on average about 40 % as much as the open funnels. The velocity of deposition to the filter pads averaged 0.4 cm s^{-1}, compared with a velocity of deposition of 0.24 cm s^{-1} to grass 34 m from the M 4.

By using the results of Cawse, it is possible to estimate the budget of emission and deposition of lead from vehicles in the U.K. (table 5). Of 8000 tonnes per year (t/a) which escapes the immediate vicinity of roads, about half is deposited over the U.K. and the remainder blown to sea. Meetham (1950) estimated that 2.3×10^6 t/a of carbonaceous material was emitted yearly in the U.K. of which 1–1.5 Mt/a were blown to sea and the rest deposited. Very similar results to those of table 5 can be obtained if the ratio of Pb (1973) to smoke (1950) in air and in deposited matter is taken on the ratio (3.5×10^{-3}) of the emissions. The current ratios of Pb to total particulates in the atmosphere is about 5×10^{-3} in rural districts (Cawse 1974) and 8×10^{-3} in London (Ball & Hume 1977).

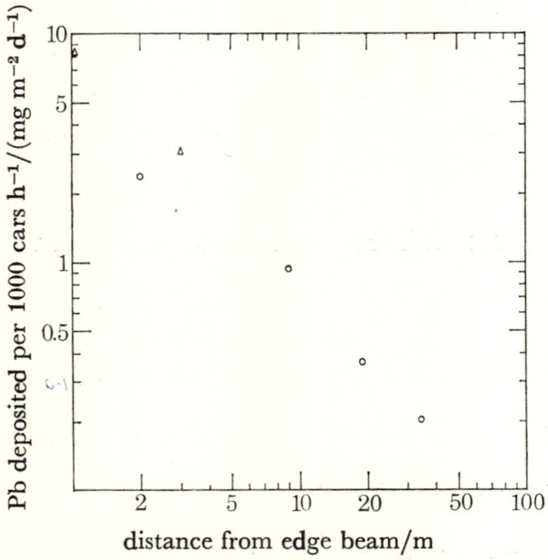

FIGURE 5. Rate of deposition of lead near motorways. Data: o, Little & Wiffen (1977); △, Huntzicker *et al.* (1975).

TABLE 5. EMISSION AND DEPOSITION OF LEAD IN THE U.K.

Emission

emission from motor vehicles (1973) (Stubbs 1972)	9000 t/a
deposited near to roads (Little & Wiffen 1978)	900 t/a
dispersed over countryside	8000 t/a

Deposition

	total area km²	rate of deposition mg m⁻² a⁻¹	deposition t a⁻¹
cities and industrial areas	5600	70	460
rest of U.K.	164000	24	3940
total deposited			4400
blown to sea			3600

5. LEAD IN HERBAGE

As well as measuring the rate of deposition of lead to freshly prepared trays of grass grown in the laboratory, Little & Wiffen (1978) and Little (personal communication) measured the lead content of grass growing on the verges and in the fields adjoining the M 4. This was done in August 1976, when grass was parched after a hot dry spell, and again in May 1977, when it was growing rapidly in a wet spring. The results (table 6) show about a sixfold difference in concentration between the two samplings. Davies & Holmes (1972), found a two- to threefold difference in concentration in samples taken near roads in Birmingham in dry and wet weather. There are two reasons for this. Much of the lead on leaves is a surface deposit removable by washing and also during rapid growth the accumulated deposit is diluted by increasing mass of herbage.

The results near the M 4, the A 491 and A 456, and the Lausanne–Geneva highway (Bovay 1970), given in table 6, all show a rapid fall-off with distance from the road. Bovay comments that it is difficult to detect the effect of emissions from a road beyond 50 m.

From table 6 it appears that the lead content of herbage at 1 m from a road carrying 1000 vehicles per hour is about 50 parts/10^6 dry matter in normal growing conditions. The corresponding rate of fallout of lead (figure 5) is about 3 mg m^{-2} d^{-1}. However, figure 5 refers to occasions when the wind is blowing from the road towards the sampling position, and the result must be divided by 2 for comparison with the data for herbage. It follows that the normalized specific concentration (Chamberlain 1970), defined as the ratio of concentration in herbage to rate of fallout, is about 35 m^2 d kg^{-1}. This compares with 24 m^2 d kg^{-1} found by Roberts *et al.* (1974) in Toronto by comparing the lead concentrations in grass with the rate of fallout of lead near a smelter.

TABLE 6. LEAD IN HERBAGE NEAR ROADS

road	cars per hour	date	10^6[Pb] in dry matter†	reference
M 4	3600	Aug. 76	876 (4), 580 (9), 428 (19), 227 (34), 122 (54)	Little & Wiffen (1978)
M 4	3600	May 77	174 (4), 108 (9), 30 (19), 26 (34), 23 (54)	Little (personal communication)
Lausanne/ Geneva	875	Aug. 69	67 (1), 42 (4), 15 (10), 19 (12), 10 (24)	Bovay (1970)
A 491 (Birmingham)	800	July/Aug.	42 (1), 19 (25), 10 (50)	Davies & Holmes (1972)
A 456 (Birmingham)	700	July/Aug.	37 (1), 28 (75), 15 (50)	

† The numbers in parentheses are the distances in metres from the kerb at which the measurement was taken.

From the results of Cawse (1974), the rate of fallout of lead in country districts of the U.K. is about 0.07 mg m^{-2} d^{-1}. Applying the same factor of 35 m^2 d kg^{-1}, this suggests that a lead concentration of about 3 parts/10^6 dry matter in herbage in country districts could be attributed to fallout. Lead in herbage harvested each year at Rothamsted since 1920 has been measured by Williams (1974). The concentrations range from 5 to 20 parts/10^6 dry matter with no significant increase over the years. There has been a significant accumulation in top soil, but sampling variations between plots gave a range of increments of 3–20 kg ha^{-1} in 50 years. This can be compared with an annual fallout of 24 mg m^{-2} (0.24 kg ha^{-1}) assumed for country districts in table 5.

REFERENCES (Chamberlain *et al.*)

Archer, A. & Barratt, R. S. 1976 *J. R. Soc. Hlth.* **4**, 173–176.
Ball, D. J. & Hume, R. 1977 *Atmos. Environ.* **11**, 1065–1073.
Bevan, M. G., Colwill, D. M. & Hogbin, L. E. 1974 *Measurements of particulate lead on the M. 4 motorway at Harlington.* Transport and Road Research Laboratory Report 626.
Bovay, E. 1970 *Mitt. Geb. Lebensmittelunters. u. Hyg.* **61**, 303–321.
Calder, K. L. 1973 *Atmos. Environ.* **7**, 863–868.
Cambray, R. S., Jefferies, D. F. & Topping, G. 1975 *An estimate of the input of atmospheric trace elements into the North Sea and the Clyde Sea (1972–3).* Report AERE-R 7733. London: H.M.S.O.
Cawse, P. A. 1974 *A survey of atmospheric trace elements in the U.K. (1972–3).* Report AERE-R 7669. London: H.M.S.O.
Cawse, P. A. 1975 *A survey of atmospheric trace elements in the U.K.: results for 1974.* Report AERE-R 8038. London: H M S.O.
Central Unit on Environmental Pollution 1975 Joint working party on lead pollution around Gravelly Hill, Interim Report, Department of the Environment.

Chamberlain, A. C. 1970 *Atmos. Environ.* **4**, 57–78.
Cholak, J., Schafer, L. J. & Yeager, D. 1968 *Am. ind. Hyg. J.* **29**, 562–568.
Cox, R. A., Derwent, R. G. & Sandalls, F. J. 1976 *Some air pollution measurements made at Harwell, Oxfordshire during 1973–1975.* Report AERE-R 8324. London: H.M.S.O.
Day, A. G., Evans, G. & Robson, L. E. 1977 *An environmental impact study of an urban motorway.* City of Bristol, Environmental Health Department.
Davies, B. E. & Holmes, P. L. 1972 *J. agric. Sci., Camb.* **29**, 479–484.
Derwent, R. G. & Stewart, H. N. M. 1973 *Atmos. Environ.* **7**, 385–401.
Fuchs, N. A. 1964 *The mechanics of aerosols.* Oxford: Pergamon.
Garland, J. A. 1977 *Proc. R. Soc. Lond.* A **354**, 245–268.
Harrison, R. M., Perry, R. & Slater, D. H. 1974 *Atmos. Environ.* **8**, 1187–1194.
Harrison, R. M. 1977 Personal communication.
Hidy, G. M., Mueller, P. K. & Tang, E. Y. 1977 *Atmos. Environ.* (In the press.)
Hogbin, L. E. & Bevan, M. G. 1976 *Measurements of particulate lead on the M. 4 motorway at Harlington, Middlesex, second report.* Transport and Road Research Laboratory Report 716.
Huntzicker, J. J., Friedlander, S. K. & Davidson, C. I. 1975 *Environ. Sci. Tech.* **9**, 448–456.
Lawther, P. J., Commins, B. T., Ellison, J. McK. & Biles, B. 1973 In *Environmental health aspects of lead*, pp. 373–387. Luxemburg: C.I.D.
Little, P. & Wiffen, R. D. 1978 *Atmos. Environ.* **12**, 1331–1341.
Meetham, A. R. 1950 *Q. Jl R. met. Soc.* **76**, 359–370.
Megaw, W. J. & Wiffen, R. D. 1963 *J. Rech. Atmos.* **1**, 114–125.
Pasquill, F. 1974 *Atmospheric diffusion*, 2nd edn. Colchester: Ellis Howrood.
Roberts, T. M., Paciga, J. J., Hutchinson, T. C., Jervis, R. E., Chattopadhyay, A., Van Loon, J. C. & Huhn, F. 1974 *Lead contamination around two secondary smelters in downtown Toronto.* University of Toronto publ. EE1.
Stubbs, R. L. 1972 In *Lead in the environment* (ed. P. Hepple), pp. 1–7. London: Institute of Petroleum.
Waller, R. E., Commins, B. T. & Lawther, P. J. 1965 *Br. J. ind. Med.* **22**, 128–138.
Whitby, K. T., Clark, W. E., Marple, V. A., Sverdrup, R. M., Sem, G. J., Willeke, K., Liu, B. Y. H. & Pui, D. Y. H. 1975 *Atmos. Environ.* **9**, 463–482.
Williams, C. 1974 *J. agric. Sci., Camb.* **82**, 189–192.

Discussion

R. M. Harrison (*Department of Environmental Sciences, University of Lancaster, Lancaster LA*1 4*YQ, U.K.*). I wish to comment upon two points arising from Dr Chamberlain's most interesting paper.

First, the question of the chemical form of the inorganic lead aerosol is important, as this may influence the efficiency with which inhaled lead is absorbed into the bloodstream. Although studies of lead in fresh vehicle exhaust have shown it to consist primarily of lead halides, little is known of the chemical form of the lead present in urban air. Our own recent studies of urban lead aerosols collected in Lancaster and Preston have shown the presence of some lead halide (PbBrCl), but the major portion of the lead has aged in the polluted urban atmosphere to form sulphates ($(NH_4)_2SO_4 \cdot PbSO_4$ and $PbSO_4$).

My second point refers to the particle size distributions reported by Dr Chamberlain. It must be borne in mind that particle size distributions determined by means of impactors are expressed in terms of aerodynamically equivalent spherical particles of unit density. These may be numerically very different from the distribution of sizes of the same particles when determined by other techniques, such as electron microscopy, as a result of non-spherical geometry, and probably more importantly, a density in the case of lead compounds of approximately 6 g cm^{-3}.

W. J. Megaw (*Department of Physics, York University,* 4700 *Keele Street, Downsview, Ontario, M*3*J* 1*P*3, *Canada*). I should like to ask if there is any evidence that any of the lead emitted by automobile exhausts is in droplet form when sampled. This is of some importance where particle size distributions are calculated from chemical analysis of impactor stages because a 1 μm droplet which has grown on a 0.1 μm particle will be collected and assessed as though it were a larger particle than 0.1 μm. Dr Chamberlain has suggested that the lead is emitted as inorganic oxides, sulphides, halides and carbonates and all of these compounds are only very slightly soluble in water (PbO, 0.0017 g/100 ml; $PbSO_4$, 0.004; PbF_2, 0.064; $PbBr_2$, 0.4; $PbCl_2$, 1.0; PbI, 0.04; $PbCO_3$, 0.0001). In spite of the high humidity in the vehicle's exhaust system one would not therefore expect that many of the lead particles would have grown into droplets before being sampled.

On the other hand, Cawse (1977, personal communication) reports that quite a high proportion (40 % in West Glamorgan, 82 % in Oxfordshire and 97 % in the Shetlands) of the lead deposited in the United Kingdom is soluble in water and he suggests that the relatively low value for West Glamorgan may be due to local resuspension of insoluble mine tailings. It may be that the lead levels are so low and the quantities of water used for solution are so large that the compounds mentioned would be classed as soluble. The other possible explanation is that the lead particles emitted by vehicles may undergo chemical change in the atmosphere and I would ask if there is any evidence for this. I would suggest that it is of some importance to determine whether appreciable quantities of lead iodide are emitted or produced subsequently as this is an effective ice nucleator and may contribute to man-made climatic change.

A. C. Chamberlain. It is correct that particle size distributions determined by impaction are in terms of mass median equivalent diameters. The density and shape of the particle affect the calibration. There is some evidence of a shift to higher m.m.e.d. values for lead in London in foggy weather, which might be caused by droplet formation on the larger particles or by more coagulation in stable atmospheric conditions.

It is unlikely that droplets would form on the primary particles of about 0.02 μm diameter. In any case, small droplets dry out in the thermal precipitator used to sample for the electron microscope. Density does not affect the particle size deduced from the diffusion box.

When exhaust aerosols are aged, halides are converted to oxides, carbonates and sulphates (Ter Haar & Bayard 1971). Bromine is lost from the particles more rapidly than chlorine (Winchester, Zoller, Duce & Benson 1967). Iodide is not emitted, and on thermodynamic grounds it is unlikely to be formed.

As the fallout of lead in country districts is about 24 mg m^{-2} a^{-1}, and rainfall typically 500–1000 l m^{-2} a^{-1}, the concentration of lead in water would only be about 20–50 μg l^{-1} if all the lead fell in rain, and this is low compared with the solubilities of the lead compounds.

The uptake of lead aerosols from the lung, in unpublished human experiments with the use of ^{203}Pb, has been found to be little different whether the subject inhaled fresh automobile exhaust, exhaust aged in sunlight, or aerosols of lead oxide or nitrate.

References

Ter Haar, G. L. & Bayard, M. A. 1971 *Nature, Lond.* **232**, 553–554.
Winchester, J. W., Zoller, W. H., Duce, R. A. & Benson, C. S. 1967 *Atmos. Environ.* **1**, 105–119.

J. D. Butler (*Department of Chemistry, University of Aston, Birmingham, U.K.*). In our investigation of airborne particulate lead concentrations at the M6–A38(M) Midlands motorway interchange mentioned by Dr Chamberlain, we have observed that the difference between the daytime and night-time concentrations decrease with distance from the junction. This is not unexpected, of course, since the traffic during the day is heavier than at night. However, when expressed in the form of a percentage of night concentration/day concentration extending over many months, the result can be used to measure the effect of traffic on air quality in the vicinity of the source.

Comparison of kerbside night/day concentrations with those at 45, 510 and 580 m gives values of 50, 61, 98 and 100 %, respectively. In other words, the day and night values cannot be distinguished beyond about 580 m because they have blended into the general urban background value of around 0.8 μg Pb m^{-3}. This is the distance, therefore, over which the interchange exerts its principal influence on air quality.

Phil. Trans. R. Soc. Lond. A. **290**, 591–607 (1979)
Printed in Great Britain

Targets of pollutants in the atmosphere

By M. W. Holdgate†
Departments of the Environment and of Transport,
2 Marsham Street, London SW1P 3EB, U.K.

People, animals, plants, ecological systems and human artefacts are all potential targets (or receptors) for pollutants. The amount of damage caused can generally be related to the dose received (defined as the product of concentration and time of exposure), and the properties of the receptor that govern uptake of the pollutant and susceptibility to it.

Plants and animals vary widely in their sensitivity to gaseous pollutants and to particulates, aerosols, and substances deposited in rain. Atmospheric pollutants may affect targets directly, or indirectly through changes in climate and, after deposition, via aquatic systems and soils. In predicting potential damage to targets each substance and type of receptor needs separate evaluation, and what usually matters is the total dose of a substance received, over what period, and in the presence of what other factors, rather than the particular proportion transmitted via the atmosphere.

Introduction

Pollutants provoke concern and social action because of the damage they are considered likely to cause. 'Damage' includes all adverse effects on man, his crops and livestock, his artefacts and his environment (Saunders 1976). It may be *acute*, resulting from short exposures to high concentrations of pollutants, or *chronic*, caused by exposures to lower concentrations over longer periods. Acute damage usually results from the near-instantaneous and irreversible response of the target or receptor, and is clear-cut: the causative agent is not too difficult to determine and controls can also be fairly direct. Chronic damage is much less easy to assess: it may involve the combination of many factors and result from sources of pollution that are distant in space and time: it may also be temporary or reversible to various degrees.

Atmospheric pollutants may affect targets or receptors (the terms are used interchangeably in this paper):

(*a*) *directly*, by (i) chemical effects on organisms; (ii) chemical effects on artefacts (corrosion); (iii) physical effects on organisms and artefacts (e.g. abrasion of surfaces, obstruction of stomata, and soiling of buildings by particles);

(*b*) *indirectly*, by (i) chemical effects on habitats (e.g. changes in acidity of soil or water); (ii) physical effects on habitats (e.g. temperature changes, changes in climate).

It is not unusual for a substance to have both direct and indirect effects. Sulphur dioxide, for example, is directly toxic to plant tissues, but may also exert an indirect effect when deposited as sulphate by wet deposition, dissolved in rain, on to soil or fresh waters. In considering the likely effect of atmospheric pollutants on targets it is therefore necessary to consider the whole series of pathways that they may follow.

† The views expressed in this paper are those of the author and not necessarily those of the Departments of Environment and Transport.

Many substances that we regard as atmospheric pollutants because their concentrations have been elevated by man are also natural components of the air. The proportion of natural and man-made sources in any particular area clearly needs to be considered in any analysis. So must the interactions between atmospheric pollutants, and the changes in ecological balance that result from their uneven impact on living targets.

In this paper no attempt is made to catalogue all the possible direct and indirect effects of atmospheric pollutants acting singly or in combination against the millions of different potential targets in the living world. Illustrations are given of some of the more important types of effects, drawing heavily on papers presented to the Study Group on Pollution in the Atmosphere over the past two years. Effects on plants receive more emphasis than those on animals because more information is available about them. Effects on the atmosphere itself are excluded because they are reviewed in other papers: in a sense, ozone in the stratosphere can be regarded as a 'target' for chlorofluorocarbons, climate and polar ice-caps as 'targets' affected by carbon dioxide, and biogeochemical cycles as targets of almost anything, but it would broaden (and weaken) the discussion too much to pursue such ramifications here.

Exposure, dose, uptake and effect

The *exposure* of a target to a pollutant is defined in terms of the concentration in the medium (or media) in contact with it at a particular time. *Dose* is defined as the total amount of the pollutant received by the target over a period. Exposures commonly fluctuate with time, many pollutants reaching short-lived peaks determined by climatic factors that affect dispersion (e.g. temperature inversions), or by sunlight, affecting photochemical reactions (figure 1). Damage commonly increases markedly with exposure concentrations, even when exposure times are short. Ozone at a concentration of 10 parts/10^8 can cause acute injury to leaves of potato (*Solanum tuberosum*) after 4 h exposure (Brennan, Leone & Daines 1964), while white pine (*Pinus strobus*) needles are damaged after the same period of exposure to only 6.5 parts/10^8 (Berry & Ripperton 1963). In contrast, exposures of pinto bean (*Phaseolus vulgaris*), radish (*Raphanus sativus*) and the pines *Pinus elliotti* and *P. taeda* to ozone at 5 parts/10^8 over 3 days, 5 weeks and 18 weeks respectively caused no visible leaf injury but resulted in chronic effects such as reduction in leaf growth, fresh mass and photosynthesis (data from Engle & Gabelman 1967; Tingey, Heck & Reinert 1971; Barnes 1972; summarized by Bell 1974). Ashenden & Mansfield (1977) review evidence that SO_2 generally does not cause visible injury at concentrations below 25 parts/10^8, but may depress growth in sensitive plants like S 23 ryegrass exposed to 15 parts/10^8 for 77 days. The time over which the dose is delivered is obviously critical, and a particular effect may *not* be produced by half the exposure over twice the time, even though the cumulative doses are identical.

The concepts of *dose commitment* and *harm commitment* (see, for example, Machta 1976) developed originally for radiation protection purposes, may be applied to other pollution problems (M.A.R.C. 1977). They imply a further step. Dose commitment is the predicted dose that a target population or populations will receive assuming a specified emission of pollutant over a period and allowing for the various factors that govern dispersion and transformation along pathways (table 1). Harm commitment is a statement of the damage that is predicted to result from that dose. As Machta (1976) emphasises, there are obvious dangers in predictions of this kind, involving extrapolation from data of limited reliability, and this is also well illustrated by table 1 which includes doubtful assumptions about the impacts of aircraft, fertilizers and chlorofluorocarbons. Even when dose commitment can be defined, it cannot be translated into harm commitment without knowledge of the interactions that occur at and within the the receptor. These interactions begin at the surface of the receptor and depend critically on its composition and on the nature of the pollutant and whether it is gaseous, particulate, a liquid droplet, or a solution in such a droplet.

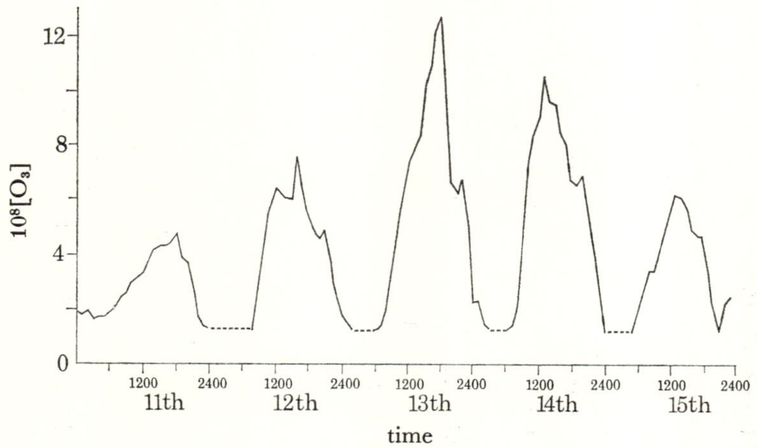

FIGURE 1. Hourly mean ozone concentrations at Silwood Park, 11–15 July 1972. Note that 1.3 parts/10^8 is the lowest value recorded because the zero of the meter was offset during calibration. From Bell & Cox (1975), with permission.

The chemistry of corrosion of many artefacts by atmospheric pollutants is well established: for example, the tarnishing of silver by hydrogen sulphide, corrosion of limestone by sulphur dioxide and acid aerosols, damage to nylon fabrics by acid aerosols, or the cracking of rubber and fading of dyes due to the ozone in photochemical smog (Jaffe 1967). Means of protection of exposed wood, metal and stone from atmospheric pollution and fungal attack have received major investment. Substitution has solved some problems: in the early 1960s, for example, stress cracking of nickel–brass alloy relays in Los Angeles telephone offices due to high concentrations of nitrates in urban dusts was solved by replacement with cupro-nickel and better humidity and ventilation controls (American Chemical Society 1969). Despite relatively good scientific understanding of the interactions, however, the costs of air pollution damage to materials was estimated at \$4.8 × 10^9 in the United States in 1968 (Council on Environmental Quality 1973), while in Britain the corrosion and protection of metals and accelerated degradation of textiles, paper, leather, etc., were together estimated to cost £75 × 10^6 p.a. in 1970 (Programmes Analysis Unit 1972).

With living organisms both the interactions and the consequent damage are much less understood. Many terrestrial plants and insects are waterproofed by surface wax layers, variously protected by layers of lipoprotein and other substances. When clean, such surfaces are not readily wetted by water and gas exchange through them is slow: in contrast, they are readily wetted by oily droplets, may be rendered permeable by abrasion by some mineral particles and adsorption by others (including carbon dusts), and may trap particles readily because they are rough. As leaves age, they often support an increasing growth of algae, fungi, yeasts and bacteria and the effects of this 'phyllosphere' on surface permeability are uncertain.

TABLE 1. ESTIMATED ENVIRONMENTAL EFFECTS (DOSE COMMITMENT) AND INCIDENCE OF AND MORTALITY FROM SKIN CANCERS (HARM COMMITMENT) DUE TO STEADY STATE RELEASE OF SELECTED CHEMICALS (FROM MACHTA 1976).

				dose commitment		harm commitment, additional cases per year†		
						incidence‡		
chemical	input 10^{11} g/a	altitude of input km	source	% ozone depletion	% increase u.v.-B	non-melanoma skin cancers	melanoma skin cancers	mortality melanoma skin cancers‡
NO_x	156	10.5	400 wide-bodied subsonic aircraft	0.082	0.12	2000	40	10
NO_x	1	16–18	100 Concorde, TU-144 aircraft	0.11–0.57	0.17–0.86	2500–15000	50–250	10–50
NO_x	164	19.5	100 large supersonic aircraft	3.27	4.91	75000	1500	300
N_2O	2000(N_2)	0	fertilizers in 2000 A.D. and constant thereafter	< 1.8–23.0	< 2.7–35.0	< 40000–500000	< 900–10000	< 150–2000
F-11§ F-12§	2 3	0	aerosols, refrigerants, etc., at 1973 rate	6.5–18.0	9.8–27.0	150000 400000	30000	600–1500

† Estimates assume a constant world light-skinned population of 10^9 persons.
‡ Based on current non-melanoma skin cancer incidence of 150/100000 light-skinned persons per year; melanoma incidence of 4/100000; melanoma mortality of 1.5/100000.
§ Chlorofluorocarbons ('freons').

In contrast to these waterproof surfaces, the stomata of green plants, the spiracles and trachea of insects, and the lungs of vertebrates are adapted for rapid gas exchange, culminating in the solution of the gas in shielded, moist surfaces.

It follows that although many pollutants, including gaseous sulphur dioxide, fluorine, nitrogen oxides and photochemical oxidants, acid aerosols, and particulate carbon, lead or fluoride may be present in polluted air, the points of contact between target organisms and

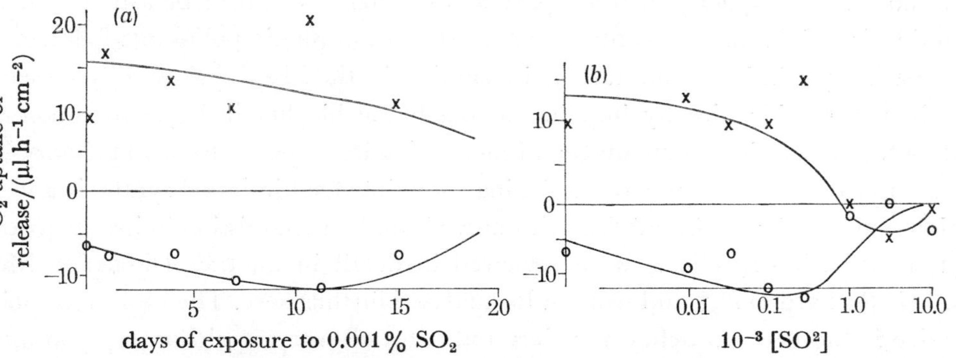

FIGURE 2. Photosynthetic activity (×) and respiratory activity (o) in disks of lichen thallus. (*a*) The effects of different duration of exposure to constant SO_2 concentration; (*b*) the response to increasing SO_2. From Ramade (1977).

pollutants and the rates of pollutant uptake must vary widely. Gaseous pollutants reach plant and animal surfaces at a rate dependent on the resistance to diffusion within the boundary layer, which in turn depends on the velocity and turbulence of the airflow over the surface (Ashenden & Mansfield 1977). Penetration within plant and animal tissues is generally fastest at sites of respiratory and photosynthetic exchange, and is greatly affected by the rates of lung and insect tracheal ventilation and on whether stomata or spiracles are open: these factors may themselves be influenced by exposure concentrations. Pollutants deposited on surfaces by wet deposition may well penetrate more slowly (pesticides and perhaps fluorine are exceptions) and may largely affect surface layers and superficial organisms like those of the phyllosphere. Particles may physically impede stomata and reduce the efficiency of animal respiratory exchange, but except where they have damaging mechanical effects, may be of more significance to animal herbivores than to plants on which they are deposited: this is certainly true of fluoride and lead particles. There are, however, some indications of interactions between dusts deposited on tree leaves and sulphur dioxide. Particles of soluble materials (like lead sulphate) naturally form an intermediate category, moving under physical diffusion in an airflow (for example within a lung) and then through solution at the tissue surface. More research is needed to document more precisely the sites and rates of entry of atmospheric pollutants into organisms.

Within target organisms, pollutants follow pathways (usually in circulating or diffusing fluids) to reach the biochemical systems on which they exert their primary effects. Oxides of nitrogen, SO_2, peroxyacetyl nitrates (PAN) and ozone all damage the mesophyll tissues of plant leaves, and SO_2 damage is manifest in the disruption of chloroplasts and depression of photosynthesis (Ramade 1977; figure 2). Ozone causes the palisade cells of the leaf to collapse through some unknown mechanism: there is some evidence that it affects the permeability of cell membranes (Bell 1976). In all these instances the pathways of the pollutant within the target are short, involving gaseous diffusion through the airspaces of the leaf and movement in solution in the cytoplasm.

In animals, too, sulphur oxides, ozone and PAN appear to act fairly directly on the cells of the mucous membranes in the respiratory tract and other sensitive tissues (like the conjunctiva). They are irritants, and provoke mucus secretion and can consequently aggravate respiratory disorders like bronchitis, emphysema and asthma. Smoke particles also aggravate such conditions, perhaps as much through the actual physical clogging of lung surfaces as through irritation, although the capacity of some particulates (like silica dust or asbestos) to cause specific disorders is well known. In contrast, some other atmospheric pollutants follow a somewhat longer pathway within animal targets in solution in the blood (whose oxygen carrying capacity carbon monoxide directly impairs) to reach the biochemical systems upon which they operate. Organochlorine compounds and metals like lead appear to have primary lesions on a range of enzyme systems, the latter affecting those involved in blood synthesis and many other functions and the former interfering with nerve impulse transmission in insects (Moriarty 1975; Clayton 1975). These effects were reviewed in detail in another Discussion Meeting convened by Sir Richard Doll,† and will not be analysed further here. The important point for us to recognize is the need to define the sites and rates of entry of airborne pollutants into targets, and their pathways and impacts there, before harm commitment can be calculated.

Variation in target sensitivity

It is a matter of common observation that plants and animals differ in their sensitivity to atmospheric pollutants: the widespread growth of the London plane (*Platanus × hispanica*) and the corresponding rarity of Scots pine and other conifers in the centre of the city is a direct expression of such variation. The differences in sensitivity between particular targets may reflect intrinsic genetic factors, or age and physiological condition (the latter depending on many other factors, such as nutritional state, reproductive condition, disease, parasite loading, etc.).

A number of authors have examined the range of variation between plant genotypes in their susceptibility to air pollution. Figure 3, based on work by Bell (1976) demonstrates that the tobacco strain Bel W-3 is much more sensitive to ozone than the strains Bel-C and Bel-B. Table 2 repeats well known data on the relative susceptibility of lichen species to SO_2 (Hawkesworth & Rose 1976; Ferry, Baddeley & Hawkesworth 1973). Within *Homo sapiens* the average annual incidence rate of malignant melanoma (which can be induced by u.v.-B radiation, and is hence one of the causes of concern over possible depletion of stratospheric ozone) varies in the United States from 4.6 and 4.4 per 10^5 (age-adjusted data) in white males and females respectively to only 0.9 and 0.7 per 10^5 in black males and females. (C.I.S.C. 1976).

These genetic differences are not necessarily static. Bell & Mudd (1976) demonstrated that in an area of Lancashire with a long history of coal-smoke pollution, ryegrass from a less-polluted coastal locality is adversely affected by prevailing pollution conditions to which indigenous ryegrass is immune (table 3). This effect probably depends on differential response to SO_2 of the processes of chlorophyll synthesis or degradation. A similar rapid selection of tolerant plant

† 'Long-term hazards to man from man-made chemicals in the environment', 13 and 14 December 1977. To be published in 1979 in *Proceedings*, series B.

strains is well known in areas highly polluted by metals (Roberts 1976; Wainwright & Woolhouse 1975).

There are many examples of variations in target susceptibility with age and physiological condition. Cooke (1970) has shown that frog tadpoles dosed with DDT in early life show a marked initial mortality, but then no further deaths until the time of metamorphosis when mobilization of body reserves brings the pesticide into circulation and promotes a second peak of mortality. On a larger scale, about 12000 guillemots (*Uria aalge*) died in the Irish Sea in

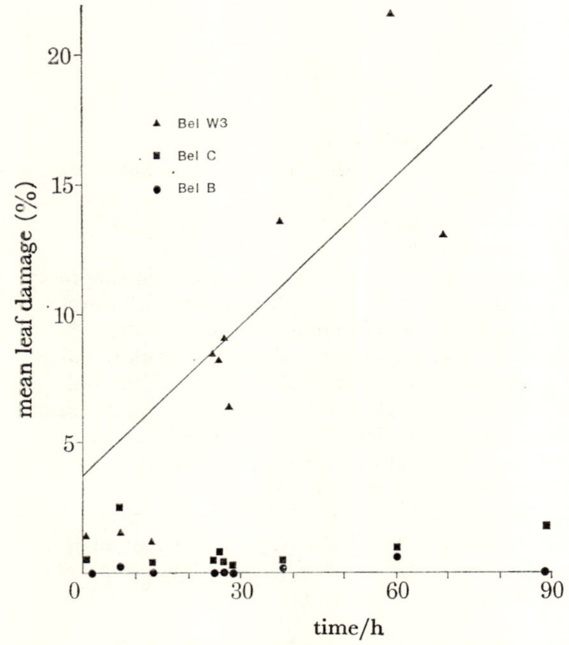

FIGURE 3. Mean leaf damage in three tobacco varieties in relation to the time that the atmospheric ozone concentration exceeded 4 parts/10^8. From Bell (1976), with permission.

1969: almost all of the dead birds were adults and the mortality probably resulted from a combination of high body burdens of polychlorinated biphenyls (p.c.bs) and strain imposed by storms at a time when the birds had already been stressed by breeding and by the moult which follows breeding (Holdgate 1971; Moriarty 1975). In man, it is well known that susceptibility to urban smogs depends very much on physiological condition: of the 3500–4000 excess deaths in London in 1952/3, most were among the elderly with histories of bronchitis, emphysema or cardiac weakness (Royal College of Physicians 1970), but the death rate of children in the first year of life also rose abruptly. Lead is another pollutant whose uptake and effect in man vary with age. Lead compounds are abortifacients and women working with lead have a high miscarriage rate, perhaps because lead passes the placenta and accumulates in the foetus throughout pregnancy (Clayton 1975). Children appear to take up more lead from the gut (perhaps because of dietary differences) and may be more susceptible to its neurological effects, although there are major uncertainties here.

A rigorous basis for comparing the susceptibility of targets to pollutant damage is often lacking. This arises because of the capacity of organisms for physiological homeostasis and reproductive and ecological adaptation, and because effects are manifest at many 'levels'. The primary lesion of any pollutant on a target organism is commonly biochemical: through the perturbation of enzyme systems and consequent effects on nerve impulse transmission, membrane permeabilities, or the rates of synthesis or breakdown of materials. We lack detailed knowledge of many of these effects, and they are hard to quantify and use as a basis for interspecific or interorganism comparison. Consequently many of the stated 'measured effects' of pollutants on organisms are at a more general level, involving measurements of gross physio-

TABLE 2. RELATION BETWEEN WINTER SULPHUR DIOXIDE CONCENTRATIONS AND THE CORTICOLOUS LICHEN FLORA OF TREE TRUNKS WITH MODERATELY ACID BARK (FROM HAWKESWORTH & ROSE 1976)

zone	epiphyte flora	mean winter SO_2 $\mu g\ m^{-3}$
0	none	?
1	*Pleurococcus viridis* s.l. confined to the base	> 170
2	*Pleurococcus viridis* s.l. extends up the trunk: *Lecanora conizaeoides* confined to the bases	about 150
3	*Lecanora conizaeoides* extends up the trunk; *Lepraria incana* becomes frequent on the bases	about 125
4	*Hypogymnia physodes* and/or *Parmelia saxatilis* or *P. sulcata* appear on bases but do not extend up trunk. *Lecidea scalaris*, *Lecanora expallens* and *Chaenotheca ferruginea* often present	about 70
5	*Hypogymnia physodes* or *P. saxatilis* extends up the trunk to 2.5 m or more. *P. glabratula*, *P. subrudecta*, *Parmeliopsis ambigua* and *Lecanora chlarotera* appear. *Calicium viride*, *Lepraria candelaris* and *Pertusaria amara* may occur. *Ramalina farinacea* and *Evernia prunastri*, if present, confined to the bases: *Plantismatia glauca* may be present on horizontal branches	about 60
6	*P. caperata* present at least on base: rich in *Pertusaria* species (e.g. *P. albescens*, *P. hymenea*) and *Parmelia* (e.g. *P. revoluta*, except in NE, *P. tiliacea*, *P. exasperatula* (in N)). *Graphis elegans* appearing; *Pseudevernia furfuracea* and *Alectoria fuscescens* present in upland areas	about 50
7	*Parmelia caperata*, *P. revoluta* (except in NE), *P. tiliacea*, *P. exasperatula* (in N) extend up trunk: *Usnea subfloridana*, *Pertusaria hemisphaerica*, *Rinodina roboris* (in S) and *Arthonia impolita* (in E) appear	about 40
8	*Usnea ceratina*, *Parmelia perlata*, or *P. reticulata* (S and W) appear: *Rinodina roboris* extends up trunk (in S); *Normandina pulchella* and *U. rubiginea* (in S) usually present	about 35
9	*Lobaria pulmonaria*, *L. amplissima*, *Pachyphiale cornae*, *Dimerella lutea* or *Usnea florida* present: if these absent, crustose flora well developed, often with more than 25 species on larger, well lit trees	under 30
10	*L. amplissima*, *L. scrobiculata*, *Sticta limbata*, *Pannaria* spp., *Usnea articulata*, *U. filipendula*, or *Teloschistes flavicans* present to locally abundant.	'pure'

A different series of indicators has been defined for trees with basic or nutrient-enriched bark (see Hawkesworth & Rose 1976).

TABLE 3. DRY MASS (grams) PER POT OF RYE GRASS (*LOLIUM PERENNE*) FROM AN AREA WITH A LONG HISTORY OF AIR POLLUTION (HELMSHORE) AND A COASTAL AREA OF LOW POLLUTION (FYLDE), AT FIRST CUT FOLLOWING THE YEAR OF ESTABLISHMENT (FROM BELL & MUDD 1976)

	locality of experiment	
locality of origin	Helmshore	Fylde
Helmshore	83.9	55.1
Fylde	68.3	105.9

logical malfunction (like diminished growth rates in plants or reduced reproductive success in birds). But these physiological effects are the resultant of many biochemical interactions, and because organisms possess a capacity to break down, eliminate or compensate for the effects of many toxic materials, there may be no manifest change at the whole organism level despite appreciable biochemical impact. For example, it is well known that the process of haem synthesis in mammals is interfered with by lead, which blocks the enzyme delta-aminolaevulinic acid dehydratase, and that this effect is measurable in man at concentrations of lead in blood above about 20 μg/100 ml, but although such concentrations are not infrequently attained within what are considered normal limits, there is no evidence of grosser impairment at the whole organism level. Similarly, even when growth rates, reproductive rates or longevity are diminished, there may be no effect on the overall population size in a target species because species possess an excess reproductive capacity. The British heron (*Ardea cinerea*) population demonstrated this in the early 1960s, when high organochlorine pesticide residue levels were associated with reduced eggshell thicknesses, causing higher breakage rates in the nest, yet although there was reduced chick survival and at the same time enhanced adult mortality, enough young were still reared to keep total numbers stable. Finally, ecosystems as a whole have a certain resilience and even when pollutant stresses reduce the abundance of some component species there may be little effect on overall appearance. In areas with mean winter sulphur dioxide levels above 170 μg m^{-3}, woodlands are likely to lack sensitive conifers and epiphytic lichens, and have only the alga *Pleurococcus viridis* on the tree trunks (Hawkesworth & Rose 1976), yet despite such impoverishment the most obvious features of the forest ecosystem would remain. Critical measurement of the nature and scale of each effect at each level in the system is thus needed if comparisons of pollutant damage are to be made.

Multiple and single impacts on targets

The classical experimental approach has been employed by many authors to investigate dose–effect relations for particular targets and air pollutants. Early work was reviewed by Webster (1967), and more recent studies on sulphur oxides by a D.O.E./N.E.R.C. Study Group (D.O.E. 1976).

This work has indicated, for example, that the ryegrass S 23, chosen as a sensitive target, can display marked yield reductions when fumigated with air containing over 50, and especially over 100, μg m^{-3}. The effect appears to be more marked in winter than in summer (Bell & Mudd 1976). When soil is deficient in sulphur, however, prolonged exposure to air containing about 50 μg SO_2 m^{-3} may favour growth, and there is increasing evidence that crops in some intensively managed areas of the English Midlands and Scottish borders would suffer sulphur deficiency in the absence of pollutant sulphur deposited from the air. There are numerous apparent inconsistencies between different experimental results, perhaps reflecting differences in technique and environmental conditions, and even the effects of other pollutants that were not strictly controlled (D.O.E. 1976). Experiments may have been conducted with closed rather than open stomata and with different rates of air movement, thus greatly modifying leaf boundary resistance and the dose received by the tissues (Mansfield 1976; Ashenden & Mansfield 1977). Extrapolation from these laboratory studies to the field, moreover, is often hampered by the use of unrealistically constant gas concentrations instead of attempting to reproduce the changes that occur over seasonal and diurnal cycles and under different weather conditions.

One way round this problem may be the use of pairs of 'open top' containers placed in the field and supplying ambient and filtered air respectively to matched crop samples: preliminary experiments have shown considerable differences in yield (Whittingham 1977).

Sulphur dioxide is rarely likely to be the sole phytotoxic pollutant present in the air. In the United States it and ozone often occur together at concentrations known to cause damage when studied separately in the laboratory. Additive, synergistic or neutralizing interactions

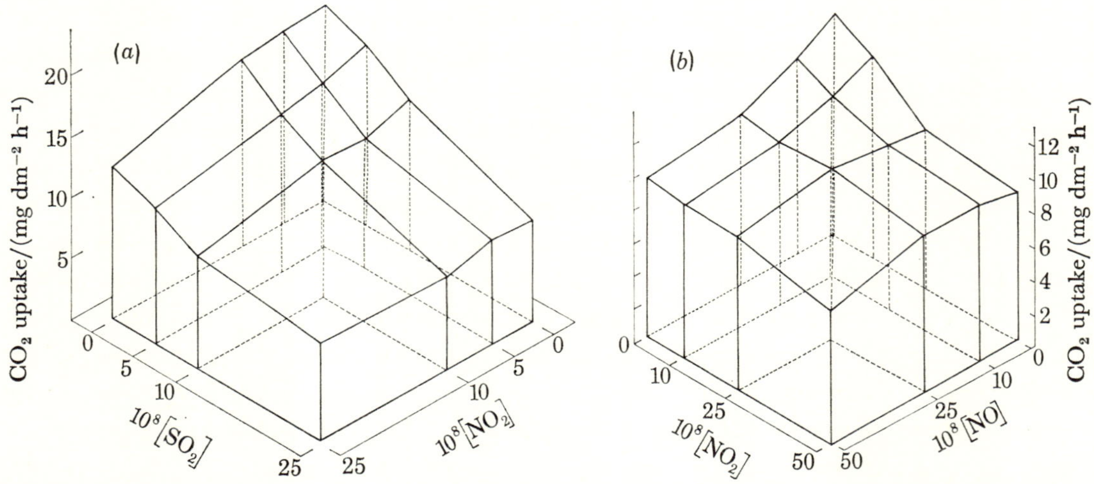

FIGURE 4. (*a*) Effects of SO_2 and NO_2 pollution on the rate of net photosynthesis in pea (*Pisum sativum*). Fumigation conditions: 1.20 air changes/min; temperature, 21 °C; light intensity, 73 J m^{-2} s^{-1}; water vapour pressure deficit 995 Pa. From Bull & Mansfield (1974). (*b*) Effects of NO and NO_2 pollution on the rate of net photosynthesis in tomato. From Capron & Mansfield (1977).

between these pollutants and others, including PAN and nitrogen oxides, consequently need to be considered. Mansfield (1974, 1977) showed that the effects of NO_2 can be additive to, or synergistic with, those of SO_2 and vice versa (figure 4*a*), and Capron & Mansfield (1976) have demonstrated a nearly additive response to NO and NO_2 (figure 4*b*). Bell (1974; also figure 3) pointed out that ozone has been shown to damage the most sensitive plants at concentrations well within the range known to occur in Britain in summer, and Menser & Heggestad (1966) demonstrated a synergistic interaction between ozone and SO_2, the two together damaging 25 and 38 % respectively of the leaves of two tobacco varieties at concentrations that were ineffective when the gases occurred singly. More recent work (J. N. B. Bell, personal communication) has indicated that depending on the target species, environmental conditions and the concentrations of SO_2 and O_3 it is possible to get a synergistic, additive or antagonistic action! Sulphur dioxide and hydrogen fluoride at high concentrations are reported as interacting additively by Saunders (1976), while ammonia and sulphur dioxide can react to neutralize one another, although the resulting ammonium sulphate haze can in special circumstances become a notorious local disamenity (e.g. as 'Teesside mist'). In case these examples appear unduly botanical, it may be worth recalling one suggestion of synergistic interaction of SO_2 and O_3 in man. Huzucha & Bates (1975) stated that healthy subjects exposed to 0.37 parts/10^6 of ozone for 2 h showed a 10 % reduction in mid-expiratory flow rate: 0.37 parts/10^6 of SO_2 had no effect over a similar period but a mixture of the two gases at the same concentration produced a more than 30 % reduction in flow rate. This effect needs further examination.

Another kind of multiple effect arises because a pollutant may affect both a target and one of its diseases or predators. Some fungi are among the plants particularly sensitive to sulphur dioxide, and rose mildew, rose black spot and maple leaf tar spot are well known to be reduced in urban areas. Oat and wheat rusts and barley powdery mildew are also sensitive to low pollutant concentrations (Bell 1976). There are also reports of pest infestations of forests after air pollution damage to foliage. Differential effects of pesticides on insect herbivores and the predators that feed on them, sometimes leading to increases in pest populations after spraying, are well known (see, for example, Moriarty 1976).

Targets are affected by many environmental variables which can alter their susceptibility to pollutants. The extent to which plants are damaged by phytotoxic gases naturally depends on exposure, which is affected by wind speed and by whether stomata are open or shut. In many plants, stomata open in the presence of SO_2, but behaviour is influenced by the water relations of the plant and there appears to be a critical relative humidity below which closure occurs. In the field bean *Phaseolus vulgaris*, radish and maize this is around 40 % while in tobacco and sunflower the critical value is nearer 50 %. Pollutants can also affect the tolerance by organisms of other variables: for example the capacity of trout to withstand cold autumn temperatures in New Brunswick was shown by Anderson (1971) to be reduced by low concentrations of DDT derived from airborne spraying of forests. Considering the complexity of the possible interactions and the inconsistency of the results obtained so far, the main conclusion to be drawn is that our understanding of pollutant interactions and the factors predisposing to sensitivity or resistance is far from adequate.

These criticisms apply especially to knowledge of the indirect affects of atmospheric pollutants. The most widely argued of these is the possibility that sulphur dioxide, oxidized and deposited as dilute sulphuric acid in rain, may be responsible for observed increases in the acidity of soils and freshwaters in some areas, with consequential changes in forest productivity and aquatic ecosystems (e.g. Sweden 1972). Research has established that long range transport of SO_2 can occur (O.E.C.D. 1977; D.O.E. 1976) and lead to wet and dry deposition at sites remote from sources, but the environmental effects of such deposition are very uncertain and there is no confirmation of any change in forest growth. Obviously, the relative contributions of local and distant sources of sulphate and hydrogen ions need to be analysed, as must the contributions to the acidification of fresh water from inorganic acids in rain, sulphate in snow melt, and humic and fulvic acids from soils. The effects of acid precipitation on soils likewise need to be compared with those of humic and fulvic acids derived from litter decomposition. Land use changes in catchment areas, climatic changes, and natural ecological trends and cycles all need to be evaluated before such indirect effects are likely to be seen in proportion.

Conclusion: the need for an integrated approach

Living targets integrate the influence upon them of many variables, and the effects of pollutants are modified by genetic, physiological and environmental factors. Three kinds of research are needed in order to unravel this complex of interactions: epidemiological studies documenting variations in target performance (and hence damage), controlled experiments to explore particular effects and mechanisms of interaction (and hence relate dose to effect) and modelling, to provide a framework for interpreting the system as a whole and guiding selection of the interactions that most require detailed study.

Some epidemiological studies have been undertaken, the most complete being that by Hawkesworth & Rose (1976) (see also Ferry *et al.* (1973) on corticolous lichens). The zones defined in table 2 can be used to predict air pollution (figure 5) and the resulting pattern correlates fairly closely with sulphur dioxide concentrations (Warren Spring 1972) but it is above all a map of industrialization. Isercutant & Margot (quoted by Ramade 1977) proposed an *Index de pureté atmospherique* (I.P.A.) on the formula:

$$\text{I.P.A.} = \tfrac{1}{100}n\left(\sum_{1}^{n} Qf\right),$$

where n is the number of lichen species present at a locality, f the frequency of each species, and Q the index of pollution tolerance for each species. A map produced by using this index is also shown in figure 5.

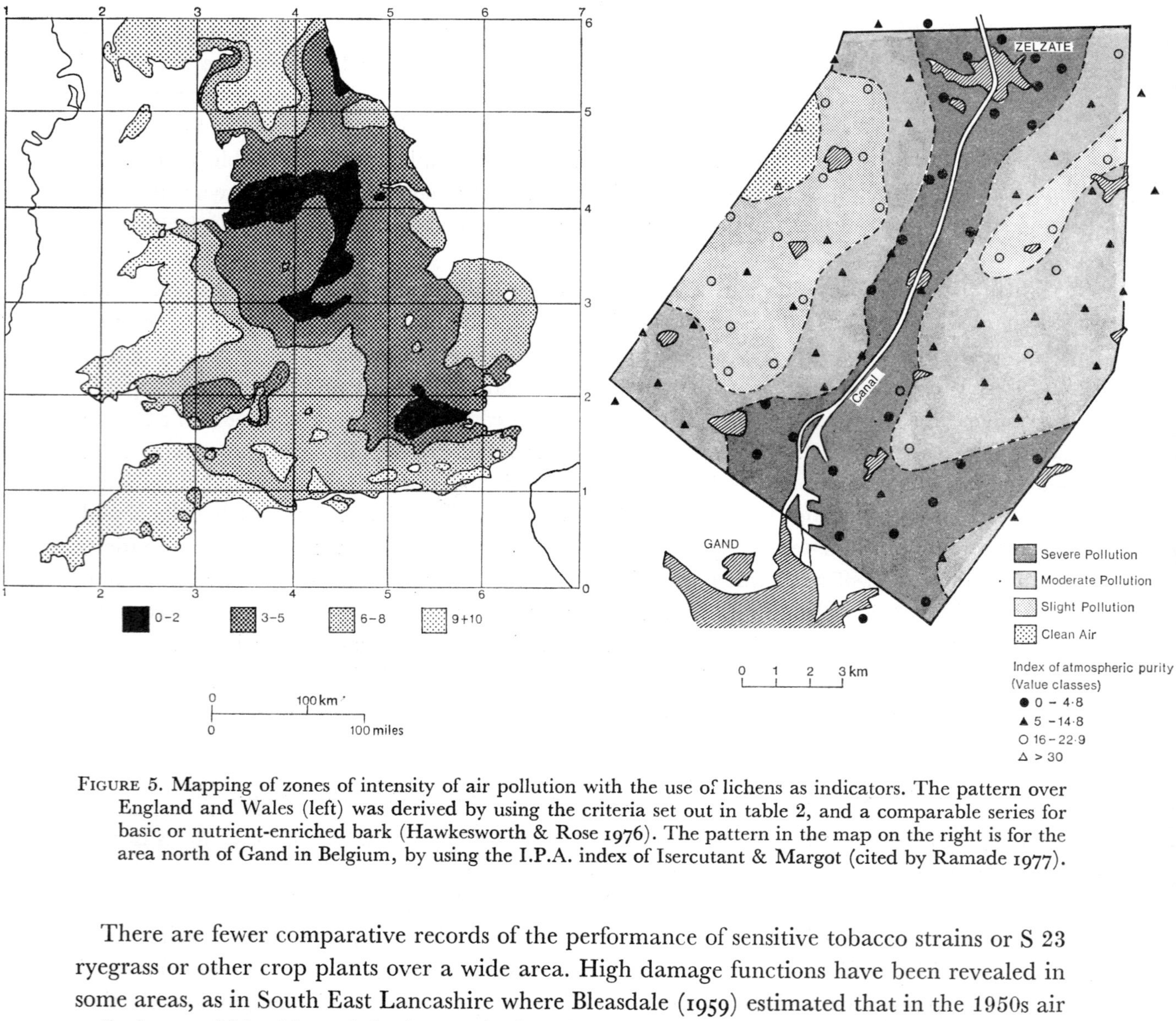

FIGURE 5. Mapping of zones of intensity of air pollution with the use of lichens as indicators. The pattern over England and Wales (left) was derived by using the criteria set out in table 2, and a comparable series for basic or nutrient-enriched bark (Hawkesworth & Rose 1976). The pattern in the map on the right is for the area north of Gand in Belgium, by using the I.P.A. index of Isercutant & Margot (cited by Ramade 1977).

There are fewer comparative records of the performance of sensitive tobacco strains or S 23 ryegrass or other crop plants over a wide area. High damage functions have been revealed in some areas, as in South East Lancashire where Bleasdale (1959) estimated that in the 1950s air pollution could be blamed for lost argicultural production worth £2.6×10^6 p.a. The lack of such epidemiological information about crop damage over the whole country may be surprising when it is recalled that the Programmes Analysis Unit (1972) provisionally estimated possible total damage costs in the region of £40 × 10^6 p.a., but may be explained in part by the practical response of farmers in not growing the most sensitive varieties in polluted areas, and by the natural development there of tolerance in pasture species. The most helpful approach may lie in the use of 'open top' containers (Whittingham 1977) to compare growth of normal crops on soil prepared by using normal husbandry methods, in filtered and ambient air. In contrast, epidemiological research on man, leading to correlations like that between bronchitic illness and air pollution (figure 6) is well established and worth extending to other relations between pollution and mortality/morbidity records. We have far less information of this kind about other animal targets.

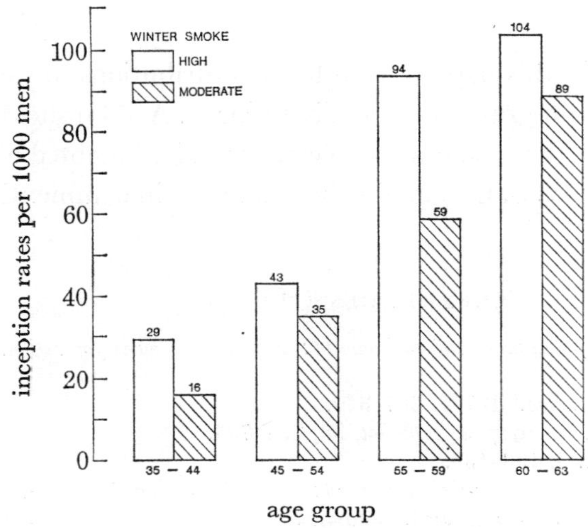

FIGURE 6. Sickness rates from bronchitis in areas in London with different pollutant levels. From Royal College of Physicians (1970), based on data from Ministry of Pensions and National Insurance Morbidity Survey.

Epidemiological studies demand establishment of pollutant concentrations and target performance with the use of sampling that permits rigorous correlation: they therefore make demands on a monitoring programme. Very often the programme proves inadequate. For example, the pattern of impaired lichen performance in figure 5 conforms with that of smoke and SO_2 pollution – but these are the only phytotoxic air pollutants extensively monitored in Britain today. It is now evident that PAN, NO_x and ozone also occur in British air, in concentrations capable of causing damage to sensitive crop plants and trees, acting in combination as well as singly, and they might also contribute to the pattern of damage, but without monitoring such relations can only be a matter of speculation.

Epidemiological surveys can only suggest relations: to define causative mechanisms experiments are essential. Classical experiments, examining dose–effect relations for one pollutant and target at a time, are necessary but are not easy to extrapolate to the field, where so many variables interact. This paper has demonstrated that laboratory research is picking its way slowly through these complex interactions, but if target performance surveys, pollutant monitoring, and experimental research are to be brought together in a synthesis, models will be required. So far, modelling has been developed for the dispersion and transformation of atmospheric pollutants, but not their impact on targets. Input–output models can be extended to the point where they define dose commitment, and it is important that this be done for all the pathways by which a pollutant reaches a target since some (lead is a good example) follow several routes and it is the total dose from all sources that matters. But models describing harm commitment need to relate pollutant dose to other environmental factors modifying target response, and allow examinination of their relative significance. It may well prove, if this is done, that air pollutants are much less significant than climate, soil, or husbandry techniques in affecting the performance of most plant targets, and that outside the most polluted urban zones they are of little concern for human health. But these hypotheses may equally be wrong, and until models allowing the exploration of the properties of pollutant–environment–target systems are built, and laboratory experiments structured to define the key component mechanisms, this field is likely to remain speculative, and one in which laboratory research is hard to relate to the real world.

Much of the information in this paper comes from contributions to the Study Group made in advance of publication. I am grateful especially to Dr T. A. Mansfield and Dr J. N. B. Bell for permission to refer to work that is still incomplete. I am also indebted to Professor F. T. Last, Dr P. J. W. Saunders and Dr W. S. Clough for their kindness in commenting on the text during preparation.

References (Holdgate)

American Chemical Society 1969 *Cleaning our environment: the chemical basis for action.* Washington: American Chemical Society.

Anderson, J. M. 1971 *Proc. R. Soc. Lond.* B **177**, 307–320.

Ashenden, T. W. & Mansfield, T. A. 1977 *J. exp. Bot.* **28**, 729–735.

Barnes, R. L. 1972 *Environ. Pollut.* **3**, 133–138.

Bell, J. N. B. 1974 In *Some gaseous pollutants in the environment: report of a seminar.* Inter Research Councils Committee on Pollution Research. London: the Research Councils.

Bell, J. N. B. 1976 Paper presented to Royal Society Study Group on Pollution in the Atmosphere on 27 April 1976.

Bell, J. N. B. 1977 Paper presented to session of N.E.R.C. Air Pollution Research Liaison Committee, Lancaster, September 1977.

Bell, J. N. B. & Cox, R. A. 1975 *Environ. Pollut.* **8**, 163–70.

Bell, J. N. B. & Mudd, C. H. 1976 In *Effects of air pollutants on plants.* Society for Experimental Biology Seminar Series, no. 1 (ed. T. A. Mansfield). Cambridge University Press.

Berry, C. R. & Ripperton, L. A. 1963 *Phytopathology* **53**, 552–557.

Bleasdale, J. K. A. 1959 In *Effects of pollution on living material* (ed. W. B. Yapp). London: Institute of Biology

Brennan, E., Leone, I. A. & Daines, R. H. 1964 *Pl. Dis. Rptr* **48**, 923–924.

Bull, J. N. & Mansfield, T. A. 1974 *Nature, Lond.* **250**, 443–444.

Capron, T. M. & Mansfield, T. A. (1977) *J. exp. Bot.* **27**, 1181–86.

C.I.S.C. 1976 *Halocarbons: environmental effects of chlorofluoromethane release.* Washington, D.C.: National Academy of Sciences.

Clayton, B. E. 1975 *Br. med. Bull.* **31**, 236–240.

Cooke, A. S. 1970 *Environ. Pollut.* **1**, 57–71.

Council on Environmental Quality 1973 *Environmental Quality: The Fourth Annual Report of the Council on Environmental Quality.* Washington, D.C.: U.S. Government Printing Office.

D.O.E. 1976 *Effects of Airborne Sulphur Compounds on Forests and Freshwaters.* A Report of the Discussion Group of the Department of the Environment and the National Environment Research Council. Pollution Paper No. 7. London: H.M.S.O.

Engle, R. L. & Gagleman, W. H. 1967 *Proc. Am. Soc. hort. Sci.* **90**, 304.

Ferry, B. W., Baddeley, M. S. & Hawkesworth, D. L. (eds) 1973 *Air pollution and lichens.* London: Athlone Press of the University of London.

Hawkesworth, D. L. & Rose, F. 1976 *Lichens as pollution monitors.* (Institute of Biology's Studies in Biology no. 66.) London: Edward Arnold.

Holdgate, M. W. (ed.) 1971 *The seabird wreck in the Irish Sea, autumn* 1969. (N.E.R.C. Publications Series C., no. 4). London: Natural Environment Research Council.

Huzucha, M. & Bates, D. V. 1975 *Nature, Lond.* **257**, 50–51.
Jaffe, L. S. 1967 *J. Air Pollut. Control Ass.* **17**, 375.
Machta, L. 1976 *The ozone depletion problem (an example of harm commitment)*. (M.A.R.C. Report no. 1.) Chelsea College, London: Monitoring and Assessment Research Centre.
Mansfield, T. A. 1974 In *Some gaseous pollutants in the environment: report of a seminar*. London: Inter Research Councils Committee on Pollution Research.
Mansfield, T. A. 1976 Paper presented to the Study Group on Pollution in the Atmosphere, 27 April 1976.
Mansfield, T. A. 1977 *M.A.F.F. tech. Bull.* (In the press.)
M.A.R.C. 1977 Report of a Workshop. Chelsea College, London: Monitoring and Assessment Research Centre.
Menser, H. A. & Heggestad, H. E. 1966 *Science, N.Y.* **153**, 424–425.
Moriarty, F. 1976 *Pollutants and animals: a factual perspective*. London: George Allen & Unwin.
O.E.C.D. 1977 *The O.E.C.D. programme on long range transport of air pollutants: summary report*. Paris: Organization for Economic Cooperation and Development.
Programmes Analysis Unit 1972 *An economic and technical appraisal of air pollution in the United Kingdom*. London: H.M.S.O.
Ramade, F. 1977 *Ecotoxicologie*. Paris, New York, Milan and Barcelona: Masson.
Roberts, T. M. 1976 Paper presented to the Study Group on Pollution in the Atmosphere, April 1976.
Royal College of Physicians 1970 *Air pollution and health*. London: Pitman.
Saunders, P. J. W. 1976 *The estimation of pollution damage*. Manchester University Press.
Sweden 1972 *Air pollution across national boundaries. The impact on the environment of sulphur in air and precipitation.* Stockholm: Royal Ministry for Foreign Affairs and Royal Ministry of Agriculture.
Tingey, D. T., Heck, W. W. & Reinert, R. A. 1971 *Proc. Am. Soc. hort. Sci.* **96**, 369–371.
Wainwright, S. J. & Woolhouse, H. W. 1975 In *The ecology of resource degradation and renewal (British Ecological Society Symposium*, no. 15) (ed. M. J. Chadwick & G. T. Goodman), pp. 231–255. Oxford: Blackwell.
Warren Spring 1972 *National survey of air pollution, 1961–71*. London: H.M.S.O.
Webster, C. C. 1967 *Effects of air pollution on plants and soil*. London: Agricultural Research Council.
Whittingham, C. P. 1977 In *Why clean air? 44th Clean Air Conference, September 1977*, Harrogate. Preprints of papers. Brighton: National Society for Clean Air.

Discussion

R. S. Scorer (*Department of Mathematics, Imperial College of Science and Technology, London SW7 2BZ, U.K.*). It is one thing, in the early stages of an industrial society, to wait for pollution to make its harmful effects apparent before restricting an industrial activity, but clearly we cannot continue that way with industry operating on the present scale. But it is quite another thing to require any theoretical hazard, however far-fetched, to be taken into account just because it has not been proved to be of no consequence. Are we not seriously in danger of this latter course? People are easily scared and seem to equate a danger which is imagined and described as a 'potential threat' as real. How do we decide whether the cry of a group that a threat is real must be taken seriously? Surely not on the basis that we must have a proof that it is not real that will convince everyone?

M. W. Holdgate. I agree that there are two philosophical starting points. One often adopted by governments, industries and others with a large stake in things as they are, assumes that the present is satisfactory until proved otherwise: changes in pollution control should depend on the demonstration of unacceptable damage or risk under present conditions. The other argues that in our present state of ignorance of environmental systems we should be cautious, never allowing air or water quality to deteriorate, prohibiting release of new substances until they have been tested exhaustively, and taking all possible opportunities to reduce pollution. In practice most Governments and agencies pursue a middle road. New chemicals are being subjected to increasingly stringent scrutiny in Europe and North America. In Britain, H.M. Alkali and Clean Air Inspectorate revise 'best practicable means' when new technology allows more effective pollution control at acceptable cost and under acceptable operating conditions. Water Authorities pursue a general policy of not allowing water quality to decline, and impose more stringent conditions on discharges when environmental needs require this or better technology permits. Pollution is, let us recognize, generally defined in damage terms. The U.N. and International Conventions state that it is 'the release by man to the environment of matter or energy liable to cause hazard to human health, damage to living resources or ecological systems, harm to amenity, or interference with legitimate uses of the environment'. There will, of course, be value judgements about the acceptability of such hazard or damage, but the fact remains that pollution is defined, and reacted to, in terms of its effects.

R. S. Scorer. In thinking about damage to the environment, which we take for granted (e.g. the air we breathe, but there are many other examples), the magnitude of the part of the environment in question is of basic importance. We can afford to put small areas of land out of action for ever, but there must be 'a big wide world out there' somewhere if the regeneration processes are to be relied upon. Consequently it is the magnitude of the human population concentrated in an area that is crucial. A city can be as large as you like if it is serviced by an appropriate area that is not built up, and this applies to each of food, mineral resources, waste disposal, etc. Do we not therefore have to develop a greater consciousness that the sheer magnitude of our population and the tonnage of its turnover have become the major cause of concern already in some areas and will soon do so in many others?

M. W. Holdgate. I do not agree with Professor Scorer that there is a direct, inevitable, relation between population and pollution. Most developed countries have experienced population growth over the past century, yet their urban air and rivers are cleaner now than they were a few decades ago. It is not the number of people that matters but the nature of their activities and the amount of money and effort they decide – and can afford – to devote to their environment. It is true that as population densities and industrialization increase we can less and less count on the dilution capacity of the environment as a 'free good' and may have to purify emissions more and more to sustain the same environmental quality. Despite some worries about relying on them too uncritically, technological solutions have so far allowed such improvement at acceptable cost. (For a fuller discussion see M. W. Holdgate (1973) in *Population and the quality of life in Britain*, London: Royal Society of Arts.)

C. P. Whittingham (*Rothamsted Experimental Station, Harpenden, Herts., U.K.*). The words 'synergistic' and 'additive' are meaningful in relation to rate measurements concerning single reaction steps. I would suggest, however, that when the parameter measured, e.g. growth, is the result of the interaction of a series of processes, each of which are inhibited by a specific pollutant to a different extent, then the use of these general terms could be misleading.

M. W. Holdgate. This is an important point. But if we are to use these terms in the strict sense Dr Whittingham suggests we may need to find new terms to apply to the interactive effects of pollutants on multi-step processes. The literature certainly mixes these two things up. In the same way, the capacity of pollutants to interact with one another and with other variables and to affect different steps in a complex process makes the concepts of dose commitment and harm commitment hard to apply unambiguously.

Elizabeth I. Young (*Fair Acres, Tydcombe Road, Warlingham, Surrey, U.K.*). In May 1976 I gave Dr Holdgate a paper reporting my research into the apparent effects of television transmission and reception during daylight on world climate. I now have a good deal of further evidence, although it is difficult to put into strict statistical form. Has Dr Holdgate been able to take any action on this matter?

M. W. Holdgate. I appreciate Mrs Young's concern, but know of no atmospheric or climatic changes that can be related to the modern increase in the use of radio waves for telecommunication: nor am I aware of any mechanism by which such an effect could come about. Indeed, as far as I know it is the properties of the ionosphere that control the telecommunication, not the other way about. Others here are, however, more qualified than I to reply.

Concluding remarks

M. W. Holdgate. The basic problem addressed in my paper is how to measure the effect of pollution quantitatively. There are so many potential targets and interactions that we cannot possibly measure them all: nor should we. Although in theory everything may be connected to everything else, it does not follow that it is good science to put equal emphasis on all these interconnections. Selection is vital, and I think there are some rational starting points. I suggest that we concentrate, at least initially, on:

(*a*) manifest effects on targets we know to be of significance (people, livestock, crops and components we have reason to believe important in environmental systems), seeking to define the scale of the effect spatially and temporally and then search for cause;

(*b*) mechanisms which we have observational or theoretical grounds for believing important, again defining the scale on which they operate, the controlling factors, and the targets involved.

Models will be needed in both fields. These need not be massive ecosystem models, but descriptions in strict mathematical language of the interactions we are concerned with, as a test of understanding, a guide to gaps in knowledge, and a means of probing relative significances of components.

Our aim, in the end, is risk assessment. We shall never eliminate risk, but we should try to define more precisely the risks we are running, or asking others to run. We can decide, in the end, to accept the risks, but we ought to know what we are doing.

Phil. Trans. R. Soc. Lond. A. **290**, 609–623 (1979)
Printed in Great Britain

The philosophy of monitoring

By J. S. S. Reay
Department of Industry, Warren Spring Laboratory, P.O. Box 20,
*Gunnels Wood Road, Stevenage SG*1 2*BX, U.K.*

The paper is concerned with the efficient design of monitoring studies on a logical basis to meet stated objectives. A series of questions are posed: why monitor, what to monitor, where to monitor, when to monitor and how to monitor?

Three broad categories of monitoring can be discerned: observation, explanation and control. The first tends to be orientated towards the receptor of pollution and the last is generally the most source-orientated. The paper discusses the requirements of different types of study. It emphasizes the desirability of monitoring the specific polluting agent having a particular effect but recognizes that sometimes a surrogate may have to be accepted. When one should measure depends on the nature of the effect produced by the pollutant in question, e.g. acute effects related to short-term peaks or cumulative effects related to long-term mean exposure. Statistical sampling should be considered but may be incompatible with some basic objectives. The siting of measuring stations is considered from the global to the local scale of monitoring and the relative attractions of mobile and stationary measuring stations are discussed. The problems of measurement of gases and particles are compared and a check list is put forward for use in selecting monitoring methods or instruments.

Methods of presenting and interpreting monitoring data are briefly discussed. Modelling of the physical dispersion of emissions is recognized as crucial to the use of monitoring data in air quality control. Much more difficult is the interpretation of biological effects with respect to pollutant exposure and concern is expressed about the tendency to attribute cause and effect relations on the basis of correlation analysis. Monitoring is not to be confused with mindless measurement.

1. Introduction

This discussion meeting is concerned with a scientific approach to air pollution problems. That in turn requires a logical and systematic approach to monitoring. It is my experience that such an attitude is not always to be found even in relation to otherwise quite sophisticated environmental studies. These studies are often carried out by specialists to whom the measurement of air pollution is peripheral. At the other end of the spectrum there are those whose elaborate proposals for scientific monitoring take little account of the very real practical constraints that exist and cannot simply be wished away.

This paper offers some observations and opinions about the overall philosophy of monitoring, trying to isolate some of the factors to be taken into account in the rational design of air pollution monitoring systems. The word monitoring itself needs some definition. It implies a variety of things to different people. There is little point in being too pedantic and accepting only one narrow definition. It is more useful to recognize the various usages but to draw a clear distinction between *monitoring* of air pollution and simply *measurement.*

The term monitoring is properly applied to the measurement of air pollution (usually ambient) over time to discern any changes produced by alterations in various factors such as emissions, weather, etc. The word has a different shade of meaning when it is applied to the measurement of air pollution (ambient concentrations, or emissions in concentration or mass terms) to demonstrate that some preset limits of quality or quantity are or are not being exceeded. Rather more questionably the term may be applied to measurements at one moment but at a number of locations to indicate spatial rather than temporal distribution. What is unacceptable is the use of the word in connection with spot measurements with no possible claim to producing representative data in time or space.

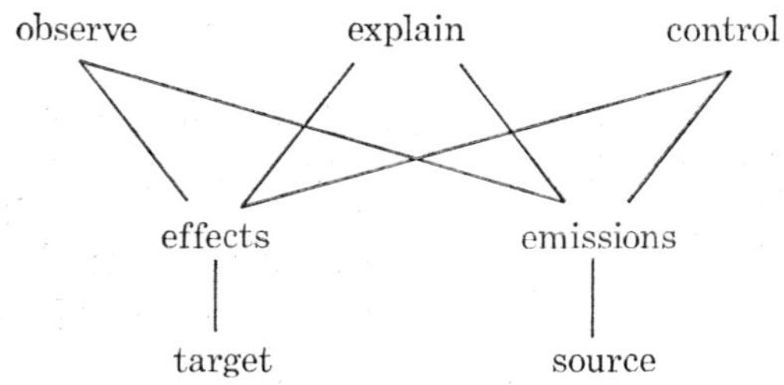

FIGURE 1. Purposes of monitoring.

2. WHY MONITOR?

The first and most important step in the logical design of an air pollution monitoring system is to set down clearly the objectives of the project. Various broad categories can be discerned.

(*a*) Monitoring may be purely speculative, looking for instance at a range of potentially harmful materials to see if there are concentrations present which should be of concern and whether there are any trends in these levels.

(*b*) Monitoring may be linked closely to studies of the effects of environmental quality factors on human health, on plants or on climate, etc., to elucidate any possible cause and effect relation.

(*c*) Monitoring of pollutants in the air or their deposition may be part of a budget assessment from source (emission) through transmission (dispersion and transformation) to sinks (immission).

(*d*) In a rather less speculative way than (*a*), monitoring may be undertaken to assess the need for emission control, e.g. around industrial sources of pollution or in streets in respect of vehicle pollutants.

(*e*) Very commonly monitoring is undertaken to provide a historical record of ambient air quality or of emissions in relation to recommended or legislated objectives or standards.

(*f*) Quite differently, monitoring may be carried out to provide the basis for real-time action to maintain emissions or air quality within regulatory limits.

(*g*) Monitoring may also be employed in relation to single or multiple sources to provide a quantitative understanding of pollutant distribution so that, for instance, the effect of developments can be reasonably predicted. This is a particularly investigative form of monitoring.

There are many other reasons for undertaking specific monitoring studies. They tend to fall into broad categories – to observe, to explain and to control. Figure 1 illustrates how various types of monitoring may be either target orientated or source orientated.

Being clear in the first place why monitoring is being undertaken goes a long way towards decisions on what to monitor, when to monitor and where, and even on how to monitor. These are not really independent questions but they can be dealt with roughly in that order.

3. What to monitor?

Why does air pollution matter? It is generally that some object or system exposed to it directly or indirectly receives a dose which has a damaging effect. Ideally, if one is interested in the relation between pollutant levels or other factors and the effects they produce, one should try to measure the effects, and the exposure to the agent or agents producing those effects. Life being what it is, however, our knowledge and our ability to measure are often so limited that we have to compromise by choosing to measure some surrogate for the effect and possibly some aspect of the pollutant which may be only indirectly related to the specific factor about that pollutant which produces the adverse effect.

For instance, records of absences from work or of deaths may have to substitute for the measurement of specific acute health effects produced by air pollution. Equally, the dirtiness of smoke and a non-specific method for determining sulphur dioxide have been used to measure the level of urban air pollution. Historically, there were good reasons for choosing to measure these things in this country in support of the Clean Air Act but we clearly recognize that it is neither the filthiness of smoke nor the concentration of sulphur dioxide in the air which is directly responsible for the effects produced by polluted urban air. For example, these measurements ignore, or are only rather tenuously related to, the presence of sulphate particles or nitrogen oxides and nitrates which may play a significant rôle in the effects of urban pollution on health.

Where possible, then, one should try to measure the precise agents and effects. Sometimes, observable effects themselves are held to be adequate indicators of the levels of specific air pollutants. Perhaps the most publicized example is the use of data on the distribution of various species of lichens to indicate levels of sulphur dioxide. This work clearly demonstrates both the advantages and the limitations of such biological indicators. Certain lichens clearly respond adversely to the presence of sulphur dioxide, either to the gas directly or to the acidity of sulphite or sulphate in water. Thus a study of the comparative occurrence of various species of lichens, whose sensitivity to sulphur pollution has been 'calibrated', may provide a relatively quick and inexpensive method of showing the distribution of such pollution over an area. It produces a useful 'snapshot'. However, sulphur dioxide is not the only pollutant to which lichens are sensitive: heavy metals and fluoride also affect them adversely. Thus lichen distribution is a clear indication of environmental quality as far as lichen health is concerned but not an unequivocal guide to levels of sulphur dioxide or to the relative effect of such environmental quality on other systems or organisms. A further limitation is uncertainty as to the extent to which the lichen distribution represents present, recent or older levels of pollution. The lichen population responds fairly rapidly to increased pollution but not so rapidly to the improvement in air quality.

Another example of an indicator is the corrosion of zinc cans. The Central Electricity Generating Boaed has carried out several country-wide surveys with the use of this relatively inexpensive technique. As a guide to the relative corrosive propensities of the atmosphere for the Board's pylons, the technique satisfies the criterion of being a measure of the precise effect in question. Although the resulting corrosion map shows considerable similarities to the distribution of sulphur dioxide levels, one can discern the effects, for instance, of sea salt and of differences in humidity or rainfall. In fact a lichen map remains a lichen map and a corrosion map a corrosion map. A sulphur dioxide map can be constructed only from sulphur dioxide measurements, emission data, and so on.

Depending on the objectives of the study in question, it is usually important to measure, or at least estimate, pollutant emissions – their quantity and spatial distribution, both laterally and vertically. Among related factors which should be considered for measurement are atmospheric conditions such as temperature, wind speed and direction, and aspects of the vertical atmospheric structure, such as mixing height and changes in wind speed and direction with height.

4. When to measure?

When, where and how to monitor constitute the subject of sampling. It covers some of the most important decisions and choices to be made in any monitoring system. First let us consider when to measure.

Whether one should measure all the time or only at intervals depends very much on what one wants from the data. Statistical sampling has substantial attractions. There are national or regional surveys in some countries where measurements at each site are made for as little as a few hours on one day in eight or even one day in twenty. In such systems it is sometimes possible to use mobile laboratories with more sophisticated equipment and more highly qualified staff than would be possible for continuous monitoring at so many locations. With such a régime it is possible to accumulate in time a picture of mean pollution levels over a wide survey area. But if, for instance, one of the prime objectives is to be able to link pollution levels with daily medical statistics, then continuous measurement is clearly required. Most air quality standards also require some form of continuous measurement. Another factor in deciding whether to measure all or part of the time is the broad temporal variability of the pollutant levels. A pilot study may have to be undertaken to clarify this before a decision is made for the survey as a whole.

This leads to the next question, which is the one of time resolution of the measurements. Here a basic criterion should be the link between the exposure to (or more precisely the dose of) the pollutant and the sort of effect which may occur or which it may be possible to record. Very probably a compromise will have to be reached, as in the U.K. National Survey. We fully recognize that short-term peaks within the mean daily values for smoke and sulphur dioxide may be important in terms of the effects they produce. Indeed, we have data to show the probable extent of short-term peaks within daily means in urban areas. If, on the other hand, a pollutant such as lead has a more or less cumulative effect on health, then it can be argued that integration over a long sampling time is a perfectly adequate form of monitoring. Figure 2 illustrates clearly the differences in the quality of data obtainable from different temporal régimes of sampling. Clearly a continuous measuring instrument with short time resolution will come quite close to registering the true variation in ambient concentrations. This may be important where as little as one gulp of a high concentration of a pollutant can be significant in producing health effects. Continuous sampling, but with coarser and coarser time resolution, obviously tells one less and less about the fine structure of exposure. A small amount of more detailed measurement may allow one to know with reasonable confidence what sort of ratio exists between short-term peak concentrations and longer-term means. Occasional sampling for only short intervals as shown in the fourth diagram really tells one very little about exposure to a pollutant whose concentration is highly variable.

If an air-quality standard is expressed in daily levels not to be exceeded more than x days per year then at least 24 h resolution of continuous measurements is required. If the standard involves a figure not to be exceeded for more than 1 h per month or year, then obviously hourly resolution is demanded. The monitoring implications of some countries' air quality standards are a not inconsiderable part of the cost of implementing them. Indeed, I would go so far as to say that the cost of monitoring should be considered in the process of defining mandatory air-quality standards.

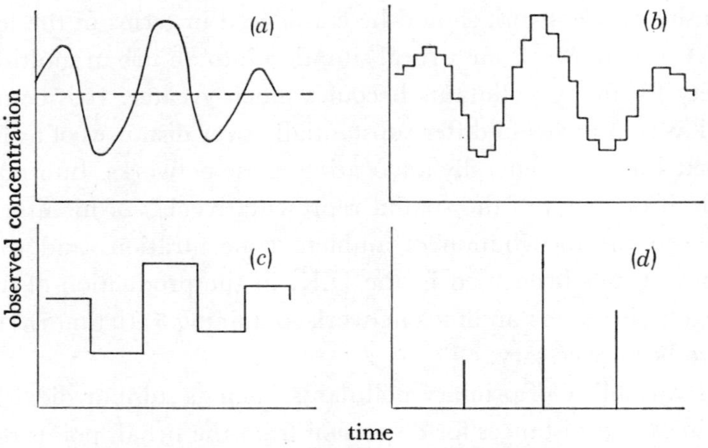

FIGURE 2. The effect of time resolution on monitoring the same air pollution event. (*a*) continuous sampling, short time resolution; (*b*) continuous sampling, medium time resolution; (*c*) continuous sampling, long time resolution; (*d*) discrete sampling, short time resolution.

5. WHERE TO SAMPLE?

On the question of where to sample air pollution, some of the major differences begin to appear between various scales of monitoring.

On the *global* scale, monitoring is usually concerned with long-term changes in background concentrations of pollutants such as particulates or carbon dioxide, which is being relentlessly produced by man's combustion of carbonaceous fuels. Alternatively, the aim may be to discern large-scale spatial variability, say between the tropics and high latitudes. Another understandable global monitoring activity is the constant watch, largely under I.A.E.A. auspices, of nuclear isotopes in precipitation. The principal siting requirement is that the places should be truly representative of background or 'baseline' conditions and not distorted by even small contributions to the levels of the relevant pollutants from specific local sources. Not all of the small number of the World Meteorological Organization's sites in this important category can match up to the basic criterion.

Elsewhere in this Discussion, reference has been made to monitoring the stratosphere. It is arguable that most of the measurements made to date do not amount to monitoring as defined in this paper. Continuous satellite measurements seem likely to provide true monitoring data.

At the *regional* level in studies, such as that carried out on long-range transport of air pollutants (L.R.T.A.P.) by O.E.C.D. countries, similar siting criteria are desirable but even more difficult to fulfil in highly developed areas such as NW Europe. In the L.R.T.A.P. project the distance and direction of 'local' pollutant sources were carefully noted for each monitoring station. If monitoring is carried out with sufficient time resolution to distinguish measurements made during different wind directions, which are also recorded, then the effects of local sources may at least be distinguished and often be allowed for.

In *national and local* surveys additional or alternative criteria may be appropriate. One may want principally to monitor the areas where pollution is highest, or likely to exceed preset limits, rather than to cover all areas. In the case of very local surveys one may wish to have sites in the area of maximum impact of an individual chimney, perhaps 10–30 stack heights from the source.

The required density of a network should be considered in terms of the level of detail and accuracy required. As one moves from a rural situation into an urban multi-emitter situation, the spatial variability for many pollutants becomes much greater. It is common for sulphur dioxide and smoke levels in towns to differ substantially over distances of as little as $\frac{1}{2}$ km. Let me say, however, that I am not generally advocating dense networks. Interpolation and extrapolation are possible if one studies the spatial representativeness of monitoring sites and the broad relation between the distribution of ambient concentrations and the distribution of emissions. This approach has been used in the U.K. in the production of national pollution maps. Without such a basic understanding a network containing 5–10 times as many monitoring locations would have been necessary.

Just as the spatial variability of primary pollutants, such as sulphur dioxide, is less in rural areas with greater times and distances for dispersion from the urban points of emission, so the spatial variability of secondary pollutants such as sulphate and ozone can be even less except where there are local removal mechanisms, such as the reaction of nitric oxide emissions with ozone, which has the effect of rendering local levels of ozone much lower.

The distribution of motor vehicle generated pollutants in streets provides a good example of very fine spatial and temporal variability which makes representative sampling very difficult. Depending on wind direction, concentrations of primary pollutants, such as CO, NO or Pb, on one side of a 'canyon' street can be ten times those on the other side. Considerable variability also exists in the vertical direction along the street. The concentrations of these pollutants at the street level are usually completely unrepresentative of conditions in the general atmosphere above roof-top level and at locations not immediately adjacent to the street. Referring back to §4, the monitoring and understanding of the levels of various pollutants at the kerbside is complicated by the fact that practically every pollutant has a different diurnal pattern depending on source configurations, local dispersion characteristics and even on atmospheric chemistry, as in the case of secondary pollutants such as nitrogen dioxide.

For modelling air pollution distribution and for other detailed studies it may be valuable to have pollution measurements at height as well as at ground level, since there may be significant gradients within the mixing layer; it may be even more important to know the height of the mixing layer itself. Particularly in meteorological conditions adverse for the dispersion of low-level emissions, high-level emissions from power stations may penetrate above the mixing layer and make little or no contribution to local ground-level concentrations. In such area studies it is necessary to have data on the differences in concentrations on hill tops and in valleys, e.g. to show the influence of plume impact.

Personal monitoring, i.e. where sampling devices are fitted to individuals, is, of course, very common in occupational hygiene work and it has the clear merit of representing accurately the exposure of a moving target. In much of our ambient monitoring we are dealing with stationary targets, in vegetation, buildings, etc., but with humans the picture is more complex. The exposure of an urban dweller is compounded of the periods he spends out of doors, possibly in significantly different parts of the area from a pollution point of view, and the time he spends indoors, at work and at home. Obviously we cannot have personal monitors on a wide scale but we should look critically at the normal practice in epidemiological work of relating the health of large populations simply to out-of-doors monitoring results. We must recognize that for most pollutants, indoor levels are much lower than those outside and therefore people spending different proportions of their time indoors and out of doors have quite different exposures. For a few pollutants, levels in certain parts of the home may exceed outdoor levels and the situation is reversed. Furthermore, over time the general habits of the population may change so that the indoor: outdoor ratio changes, and thus the relation of outdoor measurements to potential health effects may alter. There may for certain studies be a case for monitoring at selected indoor sites as well as out of doors.

6. Monitoring emissions

Where to measure as far as emissions are concerned may seem fairly simple on the face of it. A few points are worth making partly on where and partly on when to monitor emissions:

(*a*) It may not be necessary to monitor some emissions at all. For instance, sulphur dioxide emissions from fuel burning are naturally related to fuel composition and so the rate of fuel consumption can represent the rate of emission. If, however, flue gas scrubbing has to be fitted to abate SO_2 emissions, they become a potentially variable process effluent and monitoring by direct measurement may become necessary.

(*b*) Most industrial emission monitoring is decided by what is legally required: the frequency and duration of tests and whether concentration or mass emission is to be reported. Because of cost and the limited availability of techniques or robust equipment, the imposed requirements in the U.K. are usually confined to short-term measurements at quite infrequent intervals. The bulk of the measurements are made by industry itself, while the controlling Inspectorate only occasionally checks with its own independent tests. It is clearly the case that such a system cannot expect often to catch the periods of highest emissions and, human nature being what it is, measurements will sometimes be arranged to positively avoid periods of high emission. But what is the practical alternative? We cannot really contemplate an army of resident inspectors making frequent independent measurements, save in the most exceptional cases. Where there are reasonable grounds for concern, the approach to be pursued is the development of a continuous monitor, recording not merely overall emissions but giving detail over short time intervals. The Alkali Inspectorate has long indicated its attraction to a reliable 'Black Box', like an aircraft flight recorder, for problem industries.

(*c*) Remote sensing techniques are being developed which measure pollutant emissions after they leave the stack. This makes unheralded inspection of industrial emissions much more positive. It is too early to say whether the accuracy of such measurements will make them acceptable for regulatory purposes.

(*d*) Good emission data are vital to modelling studies.

7. How to monitor/sample?

There are two distinct classes of material being monitored in the air: gases and particles. They can be examined either *in situ* or on an extracted sample.

Relatively few methods exist for truly *in situ* examination, e.g. long-path spectroscopy or simple visibility measurement. Most measurement methods involve the extraction of a sample and it is here that the differences between dealing with gases and particles begin to appear.

Gases

There is not usually much difficulty in extracting a *representative* sample of air for the measurement of a gaseous constituent, at least if it is not too reactive and ephemeral.

Direct reading instruments operating on an extracted sample or gas stream can provide good time resolution of measurements although they do not quite measure instantaneous values. Where a sample is extracted continuously (or intermittently) and accumulated for subsequent analysis we begin to encounter the familiar problem of trade-offs between sampling rate, sampling time, sensitivity and cost which are dealt with in the section on selection of methods.

Particulates

It is worth expanding a little on the subject of particulate pollution. The accurate and meaningful sampling and measurement of particulate pollution in and from the air presents a difficult problem. Particulate pollution differs fundamentally from pollution by gases in that the effects of gaseous pollutants stem almost wholly from their chemical or biochemical properties, while the effects produced by particulates are often compounded of an intimate mixture of their chemical properties with their physical properties such as size, shape, colour and density. In the assessment of particulate pollution, the relation of monitoring measurements to effects is extremely complex and may vary from constituent to constituent of the particulate material present in the air.

Breathed dust

Some types of fine, insoluble and normally chemically inert particles, if respired deep into the lungs, can have serious long-term effects on health. Equally fine but soluble particles may have a quite different and more immediate effect in the lungs. If these materials subsequently get into the blood stream, their effects will depend on their toxicity in or from that medium.

Sampling of air for really fine particles is no problem but even then the sample represents the total air burden of particles in this size range and cannot itself take account of the effect of exhalation on retention, nor of such factors as the possible extent of aggregation or growth of some particles within the respiratory system which can in turn affect their aerodynamic performance in the lungs.

The inhalation of larger particles presents quite a different picture. Again some may be exhaled; some may deposit in the nose and end up in a handkerchief; others will be swallowed into the stomach. Inert particles of the same chemical constitution as those which produce serious lung effects when in fine particulate form, may produce no effects at all when swallowed but simply be excreted. Chemically reactive or soluble materials, on the other hand, may produce toxic effects by being absorbed from the digestive tract.

So we must recognize the diversity of factors involved in describing the significance of *exposure* to larger particles in air. In addition, we find substantial problems in actually *measuring* the air's content of larger particles. And the sampling problem is more difficult in ambient air subject to normal wind movements than in the relative stillness of air in most work places, which are the primary concern of the occupational hygienist.

The normal method of trapping particulates from an air sample is to collect them on a filter medium through which a known volume of air is drawn. The problem is to get the larger particles in the air to follow the path of air being sucked on to the filter. This is because of the particles' inertia. Isokinetic sampling of ambient air along the lines used in sampling emissions is too sophisticated and expensive for general use.

Work is being done and more is required on devices which will accurately sample the air for all sizes of airborne particles and, moreover, split the sample into two or more size fractions which may have different types of significance to health.

It has been mentioned that the pollution effect of airborne particles frequently involves both their physical and chemical properties. Thus characterization of a monitoring sample in terms of mass, blackness or obscuration power is not always enough. Chemical characterization may also be important. But how is this to be done? Sometimes the elemental or ionic composition of the whole sample will suffice. At times, however, the potential toxic effects may vary significantly depending on, say, the solubility of the actual compound of a heavy metal present as discrete particles in the sample. Electron scanning and other techniques are being applied with increasing precision to this problem of characterizing individual particles.

One more potential difficulty with particulates is worth mentioning. How is one to be certain that the particles in the collected sample are the same as those which were in the air? The possibility exists for the chemical and physical nature of particles caught on a filter to be subsequently changed by their being exposed to reactive gases and vapours passing through the filter or to liquid particles such as acid droplets also caught on the filter. This sort of transformation is possible especially for samples collected over long periods. For instance, a particle of chalk reacting with a sulphuric acid droplet would result in the species finally found on the filter being calcium sulphate. The problem applies also to deposit gauges.

Deposited particles

Representative sampling of particles being deposited from the air poses separate but somewhat similar problems. No deposit gauge can measure the real rate of deposition of particles of various types on to different surfaces, e.g. tarmac/grass/trees. Furthermore, different deposit gauges can show quite different efficiencies of capture for a given spectrum of particle sizes in other than calm conditions and each deposit gauge shows variability in its efficiency of capture of different sized particles with change in wind speed.

Figure 3 shows some of the results of wind-tunnel tests at Warren Spring Laboratory (C. F. Barrett & M. O. Ralph, unpublished work) on various designs of deposit gauge, the normal BS gauge (BS 1747, Part 1), the same gauge modified with a simple baffle in the receiving bowl, a simple gauge from the Norwegian Institute for Air Research (N.I.L.U.) and a rather similar design which has been proposed for consideration as an I.S.O. standard. It can be seen clearly that at a given wind speed there are large differences in the efficiency of capture for different particle sizes with each gauge and that there are substantial differences in performance between the various designs of gauge.

If ever it is suggested that there should be environmental standards for deposition rates there will not only be a problem in defining the logical basis for such standards – health or amenity –

but the choice of a standard gauge will be of real importance. Sometimes a deposit gauge is designed *not* to give a good capture over the whole range of particle sizes or wind conditions. The BS directional gauge (BS 1747, Part 5), which derives from a design by the Central Electricity Research Laboratory (C.E.R.L.), has vertical receiving slots instead of the normal horizontal receiving orifice. Thus its efficiency in a calm is theoretically zero and in winds its efficiency for capture of fine dust can be much higher than that of normal deposit gauges. This was the expressed intention of C.E.R.L., who found that the normal gauges underestimated or did not properly represent the fine deposited particulate pollution around power stations. Moreover, the recommended way of assessing the material collected in the C.E.R.L. gauge is well matched to the nature of the nuisance being studied, namely the covering of windows, walls and ledges with a fine film of dust. The particles collected in the gauge are suspended in water and the final measurement is one of the cloudiness produced in the water.

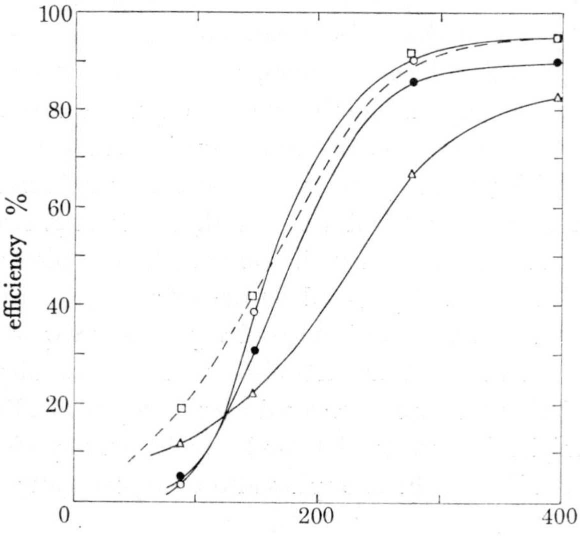

FIGURE 3. Collection efficiency of horizontal dust gauges for wind speed of 2.5 m s^{-1} at gauge height (approximately corresponds to annual mean wind speed). △, BS 1747, Part 1; □, BS modified; ○, draft I.S.O.; ●, N.I.L.U.

8. Selection of methods and instruments

To try to bring together some of the many strands involved in deciding on monitoring systems it may be useful to consider a sort of check list for use in the selection of methods and instruments for monitoring.

(*a*) *Sensitivity and limits of detection.* What is the lowest concentration or quantity of the pollutant which the method can measure with reasonable accuracy? What is the maximum concentration which the instrument reads? Is the range suitable for the situation?

(*b*) *Time resolution.* Closely linked to sensitivity. Over what sampling period or periods will the method/instrument operate? Are these consistent with the project's objectives? It is usually more expensive to measure low pollutant concentrations over short periods. A method which can be used to measure a particular level of a pollutant but requires a week-long sampling time to achieve this sensitivity is of no use if the pollution effect of interest is produced by short-term occurrence of such levels. Sometimes it may be necessary to employ two methods, one to locate and measure peaks and the other to determine the much lower longer-term mean pollution.

(*c*) *Accuracy and precision.* Are the measured results acceptably close to the real pollutant concentrations as given by a reference method and does the method give similar values on repeated measurements of the same actual concentrations of the pollutant?

(*d*) *Specificity.* Does the method measure solely the particular pollutant of interest or are the results influenced significantly by the presence of other materials in the concentrations at which they might be present in the air being sampled?

(*e*) *Reliability.* How often does the equipment need calibration or maintenance? Will it in other words operate trouble-free and accurately for long periods without attention? It is important that it should be functioning accurately throughout the period. Few people would be impressed if they bought a clock which needed winding only every three months but could well be six hours fast or six hours slow on individual days. Instruments may have automatic calibration, zero corrections and span checks but these are not provided free. Is the method or instrument sensitive to environmental conditions (heat, light, humidity, etc.) such that it might need special housing? Is it affected by voltage fluctuation?

(*f*) *Field or laboratory analysis.* Is the method one where the results are obtained, or can be obtained, at the monitoring site or must the samples be brought to a well equipped laboratory for analysis? Do the unit analysis costs fall with increasing numbers of measurements?

(*g*) *Versatility.* Can the sampling or measuring equipment be used for more than one pollutant if desired? Some methods are being developed which offer this facility.

(*h*) *Staff (operation and servicing).* What numbers and type of staff would the monitoring require and how does this compare with what can be obtained? Can an instrument be serviced by the monitoring system's own staff (possibly with the use of replacement parts) or does one have to have a complete spare unit when the faulty instrument has to be returned to the makers? How highly is the completeness of data valued? Can one tolerate gaps in data due to instrument downtime?

(*i*) *Data output.* Does an instrument provide results in a form or with a frequency compatible with the other data being collected and with the user's needs?

(*j*) *Price.* This has to be considered in relation to all the foregoing and also in relation to other factors such as what network density it will be possible to afford as a result.

It is not possible to propose a specific order for going through these check list items. In some circumstances cost and staff may be the most critical determinants; in others accuracy and specificity, for instance, may outweigh most other considerations. Then again it may be crucial to measure several pollutants concurrently. The main thing to bear in mind in considering each item is the overall objective of the monitoring. Mindless measurement is pointless if it serves no useful purpose.

9. What to do with the data

Even a summary consideration of monitoring would be seriously incomplete without something about the interpretation, presentation and utilization of monitoring data. This goes full circle back to the point at which the question – why monitor? – was asked, and in each successive section the importance has been stressed of getting the monitoring system and its results to satisfy the objectives of the study. These objectives can be so widely diverse that it is not practicable to deal with interpretation in any detail here.

It was mentioned earlier that surveys tend to be in the spectrum from observing through explaining to control in their main intentions and the position of a survey in this spectrum tends in turn to influence whether it is primarily target or source orientated, i.e. whether it is primarily interested in effects or in emissions. In principle, linkage of ambient monitoring to emissions is relatively simple (although in practice it is rarely at all simple). One is concerned with the physics which brings a fraction of the material emitted from a single source or multiple sources to be present as a particular concentration at a particular point at a certain time. One has to deal with the concepts of diffusion and dispersion, with atmospheric structure and meteorology, and to link all of these in a credible physical model of the process of transmission from emission to what our German colleagues will insist on calling immission, i.e. the receipt of the pollution by a potential target. The simplest case will be accounting for the distribution of a pollutant arising from a single emitter which is the only significant source of that pollutant. Much more complex are the cases where the pollutant arises from a multiplicity of sources distributed both laterally and vertically as in the case of urban SO_2 or smoke. Moreover, it may be important to distinguish clearly between the contributions from various source categories to provide the basis of plans for control.

One cannot control air quality simply through having quality objectives and ambient measurements. Control has to be effected on the emissions and for cost-effective control one therefore has to know a lot about how to interpret ambient monitoring data in relation to emissions. So the modelling of the physical process of dispersion is crucial. These are the easier forms of pollution phenomena to interpret and explain. The real difficulties arise when one is concerned with effects.

The effects of air pollutants on human health or living systems are rarely unique and unequivocal. Responses, as for most biological systems, will vary widely with similar exposure to a pollutant. Furthermore, there may be unmeasured variations in other environmental conditions or constituents which may also influence response. The prehistory of the exposed organism may also be important. This is only to state the obvious about all biological research. The point being raised here is a cautionary one. The interpretation of biological research data must necessarily involve statistical analysis of large numbers of pieces of information. One must not get carried away with the cult of the correlation coefficient in attributing cause and effect relations to factors seemingly linked by the existence of a high degree of correlation. There is an apposite story which Sir Edward Appleton was supposed to have told to the teetotal King Gustav of Sweden when Sir Edward was receiving his Nobel prize. This was that people get drunk on whisky and soda, brandy and soda, rum and soda, vermouth and soda and one is tempted to draw the conclusion from the data that it is soda, the highly correlated factor, which makes people drunk. The drawing of such conclusions as to cause and effect is rife in environmental studies. This is not to say that many of the conclusions are necessarily wrong. They are simply not justified by the data on their own.

While one has to be cautious in interpreting monitoring data, one should at least be as helpful as possible in presenting monitoring results. There is frequently the need to produce simple summaries of large numbers of measurements (see table 1).

In publishing summary tables for the U.K. National Survey, the general procedure has been to provide for each site the annual, summer and winter mean values and, at the monthly level, the mean value, the highest daily value and the number of days over certain levels, some of which are reckoned by the medical experts to be of significance to health. By these criteria it has been

TABLE 1. EXTRACT FROM THE ANNUAL DIGEST OF RESULTS

(Smoke and sulphur dioxide, April 1975 to March 1976; micrograms per cubic metre.)

site name and no.	class code		smoke av.	smoke h.d.	smoke no. of days exceeding 250	500	1000	2000	3000	SO_2 av.	SO_2 h.d.	SO_2 no. of days exceeding 250	500	1000	2000	3000	smoke/SO_2 ratio
Runcorn 8	A2/E	A 75	N	104						N	175						N
		M 75	28	56						84	154						0.33
		J 75			no results received							no results received					
		J 75	17	26						69	181						0.25
		A 75	21	53						63	180						0.33
		S 75	28	56						64	152						0.44
		O 75	54	155						89	168						0.61
		N 75	60	185						96	211						0.63
		D 75	71	157						95	208						0.75
		J 76	34	77						66	135						0.52
		F 76	57	126						94	211						0.61
		M 76	44	141						103	202						0.43
		mean summer	N							N							N
		mean winter	53							90							0.59
		year	N							N							N
Runcorn 9	B3/E	A 75	19	36						65	145						0.29
		M 75	N	33						N	146						N
		J 75	15	33						62	141						0.24
		J 75	14	31						41	167						0.34
		A 75	17	40						36	81						0.47
		S 75	15	41						37	96						0.41
		O 75	34	83						56	155						0.61
		N 75	52	126						74	125						0.70
		D 75	N	235						N	217						N
		J 76	N	87						N	183						N
		F 76	34	83						111	240						0.31
		M 76	N	143						N	261	1					N
		mean summer	16							50							0.32
		mean winter	44							92							0.48
		year	29							70							0.41

N, insufficient results received.

possible for local authorities to assess the priority that they should accord to smoke control or to suphur dioxide in the context of broader issues of development. Annual trend figures can be produced for individual sites, for cities, regions and for the National Survey as a whole. Care has to be taken in the selection of sites having the continuity of measurement to make their use valid in trend estimation. Interpretation of these annual trends has also to be approached cautiously since weather, for instance, can have a marked influence on the values for any one year. A run of perhaps five years is needed to eliminate the possible influence of extraneous factors.

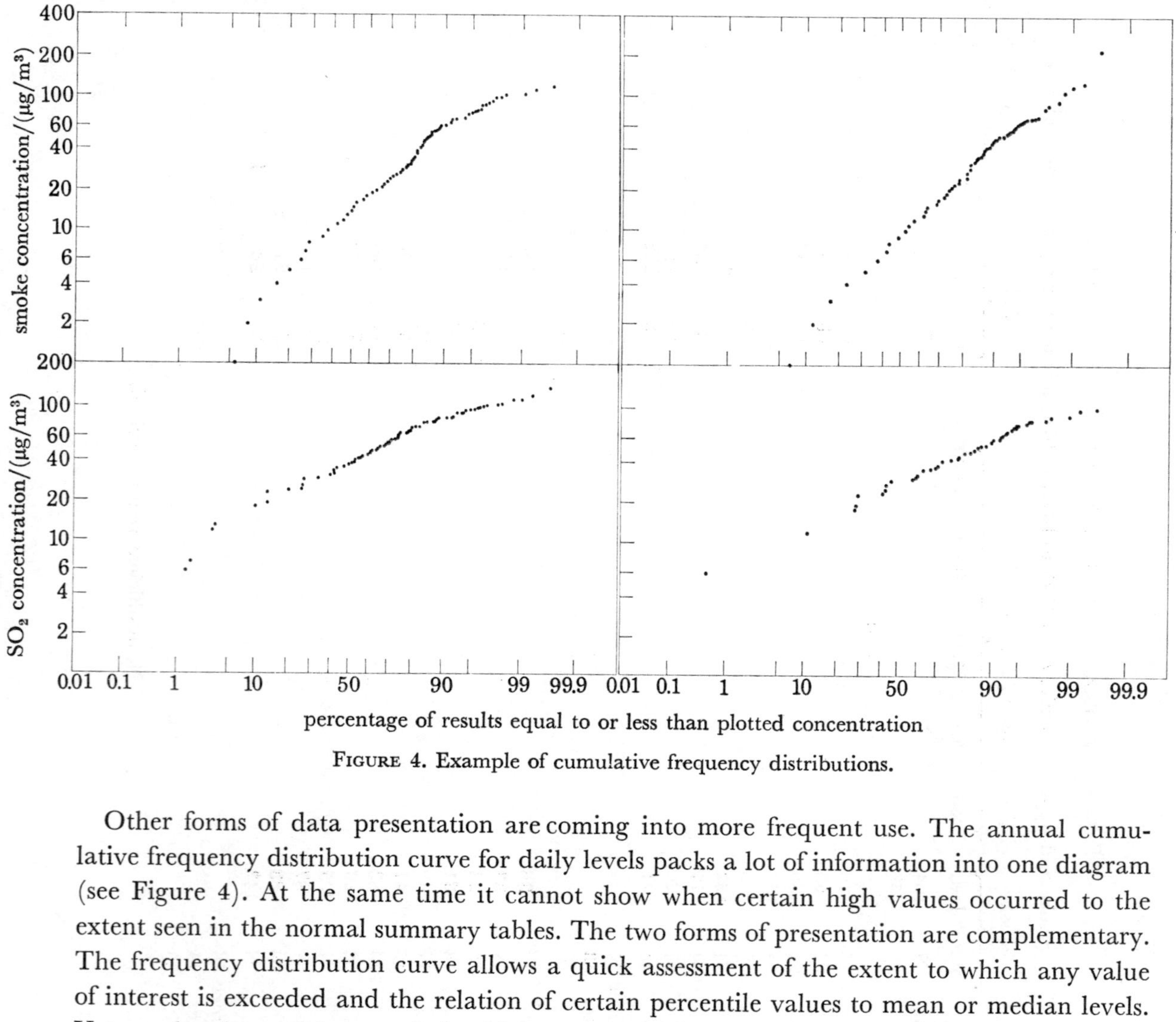

FIGURE 4. Example of cumulative frequency distributions.

Other forms of data presentation are coming into more frequent use. The annual cumulative frequency distribution curve for daily levels packs a lot of information into one diagram (see Figure 4). At the same time it cannot show when certain high values occurred to the extent seen in the normal summary tables. The two forms of presentation are complementary. The frequency distribution curve allows a quick assessment of the extent to which any value of interest is exceeded and the relation of certain percentile values to mean or median levels. Yet another form of data analysis which can be of interest from a medical point of view is the persistence of high pollution levels, e.g. the number of consecutive days when levels of smoke or sulphur dioxide, or both, exceeded particular values. Information on persistence of high pollution for two, three or more days is also important in considering the practical significance of taking temporary control measures. This sort of analysis can be readily performed on computer-stored data. With details of site location stored on the computer, computer mapping of pollution distribution is an attractive possibility. We find, however, that even with as many sites as we have in the U.K. National Survey, contour mapping cannot be reliably performed with the use of survey measurements alone. Mention was made in §5 that interpolation and extrapolation are needed and these require an understanding of the broad relation between the distribution of ambient concentrations and the distribution of emissions.

These brief references to data presentation and interpretation are purely by way of examples. It is to be hoped that they will stimulate thought on this aspect of monitoring.

More than anything else the intention of this paper is to convey the message that monitoring is not simply a hack job which can be left to a few junior staff to conduct. It is an important integral part of environmental research and control. More attention needs to be paid to designing monitoring systems to meet specific objectives, and further research and development are needed before monitoring techniques can, in fact, match the requierments of some environmental studies. At the same time there is no intention to advocate a monitoring explosion. A little good monitoring is usually worth more than a great deal of poor monitoring.

Discussion

D. T. Swift-Hook (*Central Electricity Research Laboratories, Kelvin Avenue, Leatherhead, Surrey KT22 7SE, U.K.*). As the understanding of pollution pathways improves, greater and greater sophistication is required in the measurement of more and more species. If monitoring is not to become unbearably costly, the sophisticated equipment that has to be developed must be useful for a wide range of different pollutants in different places. That philosophy is being adopted in this country.

There is a substantial programme of research at C.E.R.L. and elsewhere in the U.K. (including N.P.L. and Hull University) on remote sensing in the lower atmosphere, with some funding from the E.E.C. A major area of expertise is mobile differential absorption lidar, which enables concentrations as low as a few parts/10^9 to be measured over ranges of a few kilometres with resolutions down to a 100 m or so. By using a tunable laser, different frequencies can be scanned which cover a very large number of pollutant species with a single (albeit expensive) instrument.

We also believe that it is important to develop techniques for measurement *in situ*. Several countries, including ourselves, are becoming interested in model aircraft as a cheap way to take samples over extended periods. Not only is it desirable to complement remote scanning measurements with confirmatory spot checks, but also collection is necessary to study particulate compositions and heterogeneous reactions or catalysis that may be taking place on particle surfaces.

From the papers and discussions at this meeting it is clear that free radicals such as OH and HO_2 are becoming increasingly important. The remote laser techniques I have mentioned can, in principle, be applied to free radicals by tuning to appropriate lines. Collection techniques are much more difficult but we are attempting to apply free-radical chemistry methods used in the laboratory – cryogenic capture and subsequent spectroscopy, laser, n.m.r., e.s.r. or other methods of analysis – to air-borne sampling.

Phil. Trans. R. Soc. Lond. A. **290**, 625–637 (1979)
Printed in Great Britain

The philosophy of control of air pollution in the United Kingdom

By F. E. Ireland and D. J. Bryce
H.M. Alkali and Clean Air Inspectorate, Health and Safety Executive, Queen Anne's Chambers, 28 Broadway, London SW1H 9JU, U.K.

The United Kingdom uses an air pollution control system based on the best practicable means (b.p.m.) philosophy. The philosophy recognizes that a degree of air pollution is inevitable in present-day industrial society, and strives to optimize this pollution so that the total interest of society is best served. Account must be taken of health and amenity effects, of the capabilities of pollution arrestment techniques, of the finances of industry and effects on employment, and of the national need for the product.

This paper describes the operation by H.M. Alkali and Clean Air Inspectorate of the four basic components of the b.p.m. philosophy: the setting of emission limits and other requirements which will give acceptable levels of ambient air pollution, prior approval of works' air pollution control equipment, continuing inspection and testing of the operation of control equipment, and prosecution of works managements where requirements are not met. It is shown that the b.p.m. philosophy compares favourably with alternative philosophies based on systems of statutory air quality standards or statutory emission standards. It is suggested that the b.p.m. philosophy can be developed and, with its involvement of workers, management, local interests and control authorities, is the system best able to meet the needs of air pollution control in the future.

1. Administration of air pollution control legislation in the U.K.

(a) *Emissions from general industrial operations*

Atmospheric pollution in the U.K., other than any from traffic and nuclear installations, derives historically from two categories: general industrial operations and burning of fuels. Emissions from both categories are subject to legal control. Control is vested either in H.M. Alkali and Clean Air Inspectorate, which is a central government organization, or in Local Authorities, but the split between the two is not in terms of the industrial and fuel-burning categories mentioned. The position can be understood by first considering the situation on emissions from industrial operations, where control is applied by the Alkali Inspectorate through the Alkali Act and by Local Authorities through the Public Health Acts (see appendix).

Traditionally, the Alkali Inspectorate has been responsible for administering control of emissions to atmosphere from industries with a greater potential for pollution, perhaps because the process is new and the emission arrestment problems novel, or because the potential dangers from the emissions have only recently been recognized, or because individual plants are very large and can cause large-scale pollution. The Inspectorate strives always to solve emission control problems posed by industrial processes and for this reason seeks to assume responsibility for administering control of emissions where technical problems arise, and to relinquish responsibility where problems have been overcome. A high quality of scientific knowledge is necessary for continued control. The Alkali Inspectorate deals only with a small segment of industry, and is responsible for approximately 2200 works. Administration of control of emissions to air from the remainder of industry is the responsibility of Local Authorities; perhaps 50000 works in this category may need special attention.

(b) *Emissions from burning of fuels*

Controls on emissions from general industry derive from the situation which led up to the 1863 Alkali Act. Controls on emissions from the burning of fuels are more recent, and were precipitated by the famous London smog of 1952. The outcry at that time was followed by the setting up of the Beaver Committee, whose report (Committee on Air Pollution 1954) in turn led to the Clean Air Acts of 1956 and 1968. These Acts put restrictions on emissions of smoke, grit and dust from certain classes of boilers and furnaces, restrictions on the heights of chimneys against smoke, grit, dust, gases or fumes, and allowed establishment of smoke control areas. The Clean Air Acts are enforced by Local Authorities and apply only to premises which are not under the Alkali Act. Smoke, grit and dust emissions from industries already otherwise controlled by the Alkali Inspectorate were brought under the control of the Inspectorate in 1958, thus continuing the principle that either the Inspectorate or a Local Authority but not both should have responsibility for a given works location. The best practicable means (b.p.m.) approach of air pollution control has been evolved by the Alkali Inspectorate, and a similar approach has often been appropriate for Local Authorities.

2. The best practicable means approach to air pollution control

(a) *Basic considerations*

A report of the World Health Organization (1972) defined air pollution as 'limited to the situations in which the outdoor ambient atmosphere contains materials in concentrations which are harmful to man or his environment'. Any pollution control system must have scientific information on the effects of pollutants and their levels in the environment, and must be flexible enough to respond to emerging knowledge. Science should be concerned to bring its findings and their implications to the attention of those responsible for operation of the control system, but it is as important to know the levels of pollution that society will tolerate as to know the damage these levels will cause. Should pollution be eliminated with a probable shut down of industry, or merely abated and controlled? Society's expectation of a full life for each of its members is involved. Once basic needs for food, shelter, clothing are met, attention will turn to less pressing needs, to health care and to care for the environment and amenity. A satisfactory control system must accommodate society's expectations.

The first Alkali Act was passed in 1863 to control pollution by hydrochloric acid from the Leblanc process for sodium carbonate. Until then, the acid was emitted untreated to atmosphere. The Act required 95 % of the acid to be removed before emission of the gases to atmosphere, but it quickly became clear that the need to achieve a statutory arrestment efficiency was too rigid. Technology able to achieve lower emissions was available, yet the law did not allow these levels to be required. Hydrochloric acid discharges were limited by the 1874 Alkali Act to the more useful concentration limit of 0.2 grain/ft^3 [*ca.* 0.46 g/m^3] and the apparently innocuous requirement on alkali makers to use the best practicable means to prevent the escape of noxious gases other than hydrochloric acid was introduced. The Alkali Act was revised again in 1881, 1892 and 1906, and there have been several Orders to the Act, but no statutory limit on any emission has been added since 1881. Rather there has been an attempt to consolidate flexibility into the system by using the requirements that industries in the ambit of the Act shall use the best practicable means firstly to minimize emissions to atmosphere, and secondly to render harmless and inoffensive those emissions which do occur. This requirement is at the heart of the b.p.m. philosophy of air pollution control. The requirement never asks more than available pollution control equipment can achieve, but can always be tightened when better equipment or techniques become available.

Industrial processes change continually as time passes, new industries appear and expand, old industries become obsolescent and vanish. Flexibility in the ease with which new industries can be controlled is necessary and comes from the practice of scheduling industries under the Alkali Act rather than revising the Act itself, because changes in the list of scheduled industries are more easily accomplished than changes in the Act.

In implementing the requirements of the Alkali Act, it is the Chief Alkali and Clean Air Inspector who determines the tolerable level of pollution, and in turn what is the best practicable means of pollutant arrestment and what levels of the pollutant shall be deemed harmless and inoffensive.

The concept of practicability is still evolving. A start to understanding the principle of b.p.m. is to be found in recent legislation related to the Alkali Act. The Control of Pollution Act 1974, on noise pollution, states that '"practicable" means reasonably practicable having regard among other things to local conditions and circumstances, to the current state of technical knowledge and to the financial implications. The means to be employed include the design, installation, maintenance and manner and periods of operation of plant and machinery, and the design, construction and maintenance of buildings'. The Chief Inspector's aim is to use the best practicable means available to control pollution in such a way that an optimum blend of costs and benefits of pollution abatement is obtained for society. To do this realistically, he needs to be sensitive to society's expectations, both now and in the foreseeable future, to respond to scientific findings, to base his requirements of industry on available pollution control technology and to deal with changing industrial patterns.

(*b*) *Setting targets*

The cardinal consideration is society's expectation of the degree of air pollution which is tolerable. The target is always moving towards lower levels. Public health is the most important factor, and accordingly no emission which might cause a known health hazard can be permitted. The Chief Inspector strives to keep abreast of the latest knowledge on toxicity and aims always to achieve atmospheric concentrations which allow a wide safety margin. After health is guarded, attention must be given to amenity and aesthetic effects, and here additional complications arise in setting targets because not only will the criterion change with time, it will also vary around the country and from pollutant to pollutant. People in established industrial towns and cities are frequently more tolerant of pollution than those living in commercial or residential areas. Shall there be no damage to buildings, no smell, no visible or coloured emissions, shall grazing animals not suffer from eating polluted vegetation, etc.? Decisions here also are difficult because the relation between the costs of pollution abatement and its benefits cannot easily be established. What basis is to be used to decide whether the clouds of iron oxide fume emitted by steelworks are acceptable, or whether increasing sums of money should be spent in abatement? There is a risk that decisions may become subjective.

Recalling the definition of the word 'practicable', every decision by the Chief Inspector on pollution control requirements has to take into account the state of the work, the finances of the industry, the morale of the operatives and any likely effect on their employment, the national need for the product and the amenity of the public. A balance needs to be struck and pollution optimized so that the total interest of society is best served (Ashby 1975).

Chief Inspectors over the years have attempted to develop a system which enables the various interests to contribute to the setting of pollution control requirements. Cooperation is sought with all parties. To help maintain contact with the views of the public, Alkali Inspectors foster close relationships with Local Authority staff, take a leading rôle in local works liaison committees – whose other members are drawn from the public, from works managements and from Local Authority officials – cooperate with members of the public when complaints are made of pollution, and meet workers during routine factory visits. Equally, close collaboration is maintained with industry to find economically acceptable solutions to problems of pollutant arrestment. Working groups are set up between the Inspectorate and industry, with the latter carrying out any research and development and meeting the costs. Outside specialist bodies may be consulted, and Inspectors often travel abroad, sometimes with industry representatives, to assess foreign technology. Developments in the scientific world are always under review. Close contact between industry and the Inspectorate ensures uniformity of control demands across industry, and, most importantly for the system, ensures that the Inspectorate itself remains in touch with the latest technical developments. This involvement of the public, industry and others is a valuable feature of the b.p.m. approach because all parties then know acceptable requirements have been set and so become committed to the outcome. The Chief Inspector, however, remains responsible, and makes the final decision on all requirements. The requirements for each industry scheduled under the Alkali Act are published and easily available to interested parties.

(*c*) *Presumptive Limits*

It will be recalled that no further statutory limit on an emission has been added to the Alkali Act since 1881, yet today 60 industries with many potential pollutants are controlled. To guide works managements in the design and running of plant and to assist Inspectors in the execution of their duties, Chief Inspectors have set their own standards of emission for many of the scheduled industries. These standards represent the result which may reasonably be expected if the requirements of b.p.m. are applied satisfactorily, and are known as Presumptive Limits. Presumptive Limits have the force of law, because works which fail to meet them are presumed not to be using b.p.m. and may be prosecuted accordingly. The system is strong because Presumptive Limits can be changed easily by the Chief Inspector to meet changing pollution control needs. Presumptive Limits are uniform within an industry across the country, but can be varied to meet particular circumstances at individual works. There are scheduled works, such as coke ovens, where standards of emission cannot be set, and in these cases codes of practice may be written to help those concerned with atmospheric pollution control.

(*d*) *Prior approval*

The inspectorate maintains close contact with each industry in the broad, nationwide sense, and also seeks close involvement with developments in the individual works of each industry. This reflects the stipulation under the Alkali Act that works must obtain from the Inspectorate prior approval of the air pollution control plant before production may legally begin. Inspectors, engineers and designers cooperate fully throughout plant design and installation, and aim to develop mutual understanding of the required emission standards and of suitable means to achieve these. The Alkali Act requirement to minimize emissions is met by installation of the best practicable arrestment plant, while the requirement to render harmless and inoffensive those emissions which do occur is generally met by atmospheric dispersion from stacks sufficiently high to keep derived ground level concentrations of the critical pollutant within safe limits. The Inspectorate keeps dispersion theory and practice under scrutiny, and is alert to the possibility that pollutants for which self-purification mechanisms do not exist may cause build-up effects in the environment. Acceptable arrestment plant and, when appropriate, a stack of suitable height are integral parts of b.p.m. demands on works managements. Involvement of the Inspector from the design stage allows prior approval of the plant to be given. Inspectors do not define the precise type of plant to be used, but rather the performance to be achieved. Critics comment that this involvement with industry handicaps the Inspectorate should the plant not meet expectations, but the Inspectorate does not accept this criticism, because it is confident that plants approved are the best practicable and that if problems appear in operation every effort will be made by the company concerned to overcome them. The Inspectorate will promote and cooperate in all these efforts, but, of course, during difficulties of this kind, would never allow emissions hazardous to health to continue.

Works must also obtain from the Inspectorate approval of plant maintenance and supervision procedures, and by this means b.p.m. requirements also cover the training of operators, keeping of plant spares and standby equipment, good housekeeping, materials handling, etc.

(*e*) *Enforcement*

The close cooperation between the Inspectorate and industry in the prior-approval process has produced trust and understanding. Industry discusses its problems frankly during meetings, and in general attempts to meet its obligations towards the environment. However, the Inspectorate knows that the b.p.m. approach must be stiffened by legal sanctions against erring works. The Inspectorate is entitled to pay as many visits, unannounced, to works as may be necessary to make observations, carry out inspections and test emissions to confirm that requirements are being met. The Inspectorate frequently requires works themselves to monitor emissions to atmosphere and to record the results for examination by Inspectors on routine visits. As a general requirement, works must advise the local Inspector immediately pollutant arrestment problems arise, when production may be allowed to go ahead under conditions acceptable to the Inspector. Works which break the law are liable to be prosecuted, and Inspectors do not shrink from prosecuting when they feel it is deserved by, for instance, frequent offence, obstructionist tactics or flagrant disobedience. It is not the policy of the Inspectorate to prosecute merely because an offence has been committed by breakdown of equipment or because something has gone wrong, unless it is due to repeated carelessness, lack of maintenance or slack supervision. It is far more important to find permanent solutions to prevent a recurrence. Prosecution will not solve technical problems. The Inspectorate believes that involving industry in the problem-solving process gives the best and speediest results.

3. Alternative air pollution control philosophies

(a) *Air quality standards*

For most of the period from 1863 to the present the U.K. was free to use whichever air pollution control philosophy it chose. Now, however, the U.K. must expect increasing pressure to conform to air pollution control philosophies in use elsewhere, and it is in the interest of all that the best system be chosen. Alternative systems must be examined. Modern air pollution control philosophies are often based either on a system of air quality standards or on a system of emission standards (de Nevers 1977) and both are frequently applied, outside the U.K., on a statutory basis. The air quality standards approach has the seemingly logical premise that acceptable levels of pollution can be defined and steps then taken to ensure that actual pollution never rises above these levels, but in practice there are many difficulties (Scorer 1973). The acceptable level for each pollutant has to be established, on bases which are likely to vary from pollutant to pollutant as the critical exposed species (child or aged person, human or animal or plant, etc.) varies, as the critical mode of action (short- or long-term exposure, accumulation, etc.) varies, as the presence of synergistic effects from other pollutants varies, and so on. Precisely because the system is statutory, any standard proposed must cover each of these factors and will be complicated. The complication is likely to be such that legal enforceability must be open to doubt, but, disregarding this, there are further problems. Any law can only be enforced when information on how well the requirements are being met is available. An air-quality-standards law demands an efficient monitoring system to determine pollutant concentrations in the ambient atmosphere, and the technical and economic problems in setting up such a system are immense. In turn, if such a system could be established and then did show that the required standards were not being met, further difficulties would follow, because the statistical nature of atmosphere dispersion and practical monitoring techniques is such that it would be impossible to assign responsibility to any specific works. The probable action would be a demand that all sources decrease their emissions by a specified amount, but this is extremely inefficient because there is no good way of selecting the works which must act. Are only 'nearby' works involved, or only those 'upwind'? The method will allow little suitable action against individual sources which are emitting excessive pollution and will disregard the economic and political considerations at the various works affected by the restriction. The system seems unnecessarily complicated and impracticable.

(b) *Emission standards*

Problems arise with the air-quality-standards approach because it attempts to base action directly on events in the atmosphere at large. Pollution can only be controlled by restricting emissions to air, and so a system of emission standards must be considered. Emission standards set limits to emissions of specified pollutants from particular processes. On occasion, standards which have not related to the performance of existing pollutant arrestment plant have been set, sometimes unintentionally but at other times in reflexion of a conscious policy of establishing ambitious targets to encourage satisfactory arrestment plant to be evolved and installed. Frequently this plant has not appeared, the law in general and pollution control in particular have been brought into disrepute, and pollution has continued unabated. Many types of emission cannot be quantified and so cannot be controlled by a system of emission standards.

It will be recalled that in the early years of the Alkali Inspectorate, air pollution control suffered because the rigid demand for arrestment of 95% of the hydrochloric acid rapidly became a licence for plants to pollute up to this level although equipment able to achieve better levels was available. Systems which operate statutory limits generally risk granting licences to pollute. This remains true today of statutory air quality or emission control systems.

(*c*) *Relation of the b.p.m. approach to the air quality and emission standards approaches*

The various criticisms above of the so-called air-quality standards and emission control standards approaches are over-simplified, but they highlight real difficultes which any satisfactory air-pollution control system must overcome. The air-quality standards and the emission control standards approaches each describe only aspects of air pollution control. A truly satisfactory system must seek to combine the strengths of both, while avoiding the failings.

The objections to a statutory emission control system concern primarily the inflexibility imposed by the statutory nature of the system and the theory that over-tight requirements will be met by the appearance of more efficient plant. The b.p.m. approach works through emission control but avoids rigidity and overambition in the arrestment standards required. Again, the Chief Alkali Inspector, when he takes note of the best scientific opinion on toxicity, etc., and tempers this with what he learns in feedback from the public, industry and his various other sources of advice in deciding tolerable atmospheric pollution levels, is actually operating an air-quality system which avoids the rigidity of statutory standards. The b.p.m. system does not propose immediate enforcement action if the levels of pollutants in the ambient atmosphere exceed specified limits, and thereby does not become caught in the self-defeating events which follow such action. In the b.p.m. approach, realistic emission standards are linked to amply sufficient air quality requirements by means of sufficient stack dilution of pollutants. The b.p.m. approach does combine the strengths of the emission control and air quality control systems while avoiding the failings, and yet has additional strengths in the need for prior approval of plant, approval of maintenance and supervision procedures, training of operators, etc.

4. Implications of long-range transport of pollutants for b.p.m.

The recent findings (O.E.C.D. Programme 1977) on long-range transport of sulphur dioxide from Europe to Scandinavia may seem to call in question the b.p.m. approach because of its dependence on atmospheric dilution, but this is to overlook the fact that the approach seeks always to optimize pollution. Since the time of the 1956 Clean Air Act approximately, the pattern of domestic heating in the U.K. has shifted from open fires to indirect methods which very often have been electrically powered. The shift has lowered ambient sulphur dioxide levels in the U.K.'s urban areas and has led to improved health for the population, yet the total sulphur dioxide from the burning of fuel in the U.K. has changed little. Estimates based on the O.E.C.D. study suggest that the change from open fires to central power-stations is equivalent to an increase of 1–2% in the sulphur dioxide received in Scandinavia. Methods for lowering the sulphur dioxide content in power station exhaust gases are available but expensive, and often involve secondary problems such as the disposal of calcium sulphate. To date, the optimum cost effective application of the b.p.m. approach in the U.K. interest has been dispersal of sulphur dioxide untreated in the atmosphere. The b.p.m. approach will not have been negated if this ceases to be acceptable. There will simply have been a shift in the optimum, and b.p.m. then perhaps will entail desulphurization.

5. Uniformity in presumptive limits

Before leaving the more technical aspects of how the b.p.m. approach may be expected to meet the demands of the future, it is worth noting that there will be bounds to the size of the area over which uniformity in presumptive emission limits is possible. Presumptive limits are part of the equilibrium between the performance which arrestment equipment can reasonably attain, the atmospheric dispersion which a stack policy can achieve, and the air-quality target which must be met. Considering the wider area of the E.E.C. only, it is probable that the countries involved could agree a set of uniform air-quality targets for the important pollutants, but uniformity in pollutant arrestment, given the varying wealth of the countries, and in dispersion performance, given the climatic variations of the countries, would be difficult to achieve. Again, of course, failure to establish uniform presumptive limits across an area as large as the E.E.C. is not a negation of the b.p.m. approach, since there is no technical reason why a regional b.p.m. approach, with regions each operating a separate set of presumptive limits, could not be devised.

6. Widening the decision process in U.K. air pollution control

There is no doubt that the b.p.m. approach can arrive at the optimum solution that air pollution control requires. This is not to say that air pollution control in the U.K. does not merit criticism or that it cannot be improved, only that on examination the criticisms are not of the approach itself but of the way it is implemented. For example, outcries occasionally occur about emissions from specific works, and the claim is made that the air pollution control system has failed. Certainly this will seem so to the housewife whose washing is soiled by black particulates. Here there might have been failure by the Alkali Inspectorate correctly to assess the tolerable pollution level, or failure in the information system through which the Inspectorate assessed the tolerable pollution level initially. Again, the local employment situation or the wider national need might justify the existence of the offending works and so on. However, none of these explanations would point to failure of the b.p.m. philosophy itself.

Improvements must continually be sought in air pollution control, but it seems unlikely these will be through a philosophy other than best practicable means. They are more likely to be an evolution of the b.p.m. approach, and in the mechanics of how it operates. There is a need for greater appreciation by the public of the types of input – scientific, economic, etc. – to b.p.m., and for greater confidence by the public in the control authority's ability to weigh these inputs satisfactorily. The decisions required in air pollution control cannot avoid being political, in the non-party sense, and it may be that the public's confidence will only increase as its political involvement increases.

The public must be involved in such a way that a positive gain results. The Control of Pollution Act 1974 gives Local Authorities the right to acquire and publish information on works' emissions to air, and this is certainly to the good. Beyond this, however, the prospects are unclear. Decisions by the Government on the recommendations (Royal Commission on Environmental Pollution 1976) in the Fifth Report of the Royal Commission on Environmental Pollution are awaited. The recommendations are many: that b.p.m. should be defined precisely for individual works, that air-quality guidelines should be explicitly formulated and publicly declared, that works' plans for improved pollution control systems should be published and publicly justified, that the control authority should develop expertise in economics, etc. Many of these suggestions are clearly valuable, but it would be wrong to comment specifically in this paper. It is pertinent, however, to remind all that the need is to introduce a format for beneficial consultation without defeating the purpose by clogging the system, and that the Alkali Inspectorate already has long sought means, for instance through Local Liaison Committees, etc., effectively to involve the public in the air pollution control process.

7. Appendix. Air pollution control legislation in the United Kingdom

1. Legislation enforced by H.M. Alkali and Clean Air Inspectorate:
   The Alkali Etc. Works Regulation Act 1906.
   The Alkali Etc. Works Orders 1966 and 1971.
   The Health and Safety at Work Etc. Act 1974.
2. Legislation enforced by Local Authorities:
   The Public Health Acts 1936 and 1961.
   The Public Health (Recurring Nuisances) Act 1969.
   The Clean Air Acts 1956 and 1968.
   The Control of Pollution Act 1974.

3. H.M. Alkali and Clean Air Inspectorate became part of the Health and Safety Executive in 1975. The Executive contains bodies such as H.M. Factory Inspectorate, H.M. Mines and Quarries Inspectorate and H.M. Nuclear Installations Inspectorate, and is charged with protecting persons throughout the U.K. against hazards from work activities. The intention is that the Alkali Etc. Works Regulation Act 1906 and the 1966 and 1971 Orders will be subsumed by the Health and Safety at Work Etc. Act 1974. Section 5 of the 1974 Act replaces the main provisions of the Alkali Act and contains the philosophy of b.p.m. The section will be enforced by the Alkali Inspectorate. Section 3 of the 1974 Act is a general section dealing with the effect of work activities on the public so far as health and safety are concerned and includes air pollution from processes not scheduled under the Alkali Act. The section is enforced by the Factory Inspectorate.

4. H.M. Alkali and Clean Air Inspectorate operates only in England and Wales. Equivalent controls are operated by H.M. Industrial Pollution Inspectorate for Scotland, and by the Alkali and Radiochemical Inspectorate for Northern Ireland. These Inspectorates each use the b.p.m. philosophy.

References (Ireland & Bryce)

Ashby, Lord 1975 *Clean Air*, Winter, pp. 9–15.
Committee on Air Pollution 1954 Cmnd 9322. London: H.M.S.O.
de Nevers, N. 1977 *Air Pollut. Control Ass.* **27**, 197–205.
O.E.C.D. Programme on Long Range Transport of Air Pollutants 1977. Paris.
Royal Commission on Environmental Pollution 1976 Fifth Report, Cmnd 6371. London: H.M.S.O.
Scorer, R. S. 1973 *Pollution in the Air*. London: Routledge & Kegan Paul.
World Health Organization 1972 *Technical Report*, no. 506. Geneva.

Discussion

W. J. MEGAW (*Department of Physics, York University,* 4700 *Keele Street, Downsview, Ontario M*3*J* 1*P*3, *Canada*). While Mr Ireland and Dr Bryce have demonstrated that the 'best practicable means' approach is an effective method of emission control, I would suggest that this only addresses half the problem for it makes no attempt to prevent or control the potentially disastrous short-term air pollution episode which depends not only on the sources, but also on weather and local geography. I am not a Canadian so perhaps I may be forgiven for boasting a little about the air pollution control system operated by the Ministry of the Environment in Ontario which I believe to be one of the most effective in existence. The Ministry operates the 'best practicable means' of control of emissions both in regard to existing and proposed factories but, in addition, the 'Air Pollution Index' (a function of the sulphur dioxide and particulate content of the atmosphere) (see *Ontario's Air Pollution Index* (1977), Toronto: Ministry of the Environment) is measured continually at a large number of places in Ontario. Values of below 32 are considered acceptable but when a value of 32, combined with an unfavourable weather forecast for the next 6 h, is reached, the Minister warns specified industrial concerns that they may be required to close, further factories are closed down at an A.P.I. of 75 and at a value of 100 all sources not essential to public health and safety are required to shut down. In practical terms it is expected that people with chronic respiratory disease may experience accentuation of symptoms in SO_2 concentrations of 0.21 parts/10^6 combined with particulate concentrations of 300 $\mu g/m^3$ (Lawther, P. J. 1958 *Proc. R. Soc. Med.* **51**, 262–264) and at this level the Ontario A.P.I. would be 58. At an A.P.I. of 100 it is expected that mild effects would be experienced by healthy people and that those with severe cardiac or respiratory disease would be seriously endangered. Had the Ontario A.P.I. been measured during the London smog of 1962 (which caused a 20 % increase in mortality) its value would have been 586.

The effectiveness of the 'best practicable means' method of emission control has been demonstrated in Ontario by the progressive annual reduction in areas exposed to specified mean annual concentrations of sulphur dioxide. The episode prevention system also works. The number of times per year in which the A.P.I. in Toronto exceeds the value 32 is generally between 0 and 3. I can remember only one occasion in the last five years on which the value 50 was exceeded and some factories were shut down. The Ontario system thus prevents the build-up of the short-term episode and I should like to ask Dr Bryce whether he does not consider that such a system would be of advantage in the United Kingdom?

D. J. BRYCE. The b.p.m. approach is such that an attempt is made, when setting the presumptive emission limits for different industries, to ensure that the emissions are set at a level which makes it unlikely that pollutant concentrations in the atmosphere will ever build up to the levels needed for an episode to occur. However, industry can only reasonably be asked to meet emission requirements which will be satisfactory in weather conditions which are likely to arise. Clearly, abnormal weather conditions over large industrial areas may call for special emergency arrangements, but this is within the spirit of the b.p.m. approach. A system does exist to deal with pollution episodes in the U.K.

F. E. IRELAND. Many years ago an interdepartmental working party was set up to decide on a plan of campaign in case of another serious High Pollution Episode. The working party considered what could be done in advance to reduce the risk and also what procedure should be adopted if and when an Episode occurred. Expert opinion was that more people were at risk today because medical science kept seriously ill people alive who in the past would have died. These people needed only an extra stress to push them over the brink and among these stresses could be abnormal pollution. Further, weather conditions had not changed significantly since the 1952 and 1962 Episodes and the same conditions could recur, although the results would not be expected to be so dramatic as in the past because of the reduction of pollution. Briefly, a careful watch is kept by the Meteorological Office on weather forecasts to identify critical conditions. The continuous pollutant monitoring instruments operated by Warren Spring Laboratory and the Medical Research Council are watched for significant rises of sulphur dioxide and suspended particulates. When a High Pollution Episode is suspected, extra monitors are brought into play and certain Departments and individuals are alerted to watch the situation. If it looks like becoming serious, the public, especially those predisposed to bronchial trouble, are given advice through the media. If an advanced stage were reached, by consultation some industries would be asked to moderate their production programmes. The situation would be treated in a flexible manner by discussion between the small team of experts involved.

B. PAGE (*Science Division, Polytechnic of Central London,* 115 *New Cavendish Street, London, W.* 1, *U.K.*). Although the Alkali Inspectorate is responsible for the industries thought to constitute a great pollution risk, the majority of industrial premises in the U.K. are unregistered. I became aware of this problem when I was recently involved in a survey which revealed air levels of up to 11 μg cadmium/m^3 air outside a London factory. Cadmium levels of up to 387 parts/10^6 were found in household dust from nearby flats. The pollution had apparently been occurring for the last five years, and had the survey not been carried out, it would presumably still be occurring now.

Many small firms, such as scrap metal companies, may be reticent about bearing the cost of expensive equipment to contain invisible pollutants. The local authority officers have no legal standards against which to compare the potential danger of the pollution, and may be hampered in their action. At what level in the environment does cadmium, for example, constitute a health hazard on which they should act?

Perhaps more important is the question of how such incidents of pollution can be discovered in the first place. Pollutants are so diverse, and sampling techniques so difficult and therefore expensive, that local authorities do not have the facilities to monitor emissions from the tens, possibly hundreds of thousands of unregistered premises they are mandated to control. Such unregistered premises may be situated in the heart of residential areas. Pollution is discovered when:

(1) A survey for the correct pollutant in the correct locality is undertaken. A simultaneous survey for lead in the area polluted by cadmium described above indicated that pollution levels were not abnormally high for such an urban area.

(2) The pollution reaches a sufficiently high level for it to become noticeable to local people. Noise attracts many complaints, and levels are often rigorously enforced. Unfortunately, even the high levels of the toxic metal cadmium remained invisible to the local community and factory workers throughout the 5-year period of pollution. Since cadmium has a biological half-life of up to 30 years, cessation of production does not immediately eliminate their problem.

The chances of early discovery of toxic pollutants from Britain's unregistered industries are not good. In the absence of a coordinated monitoring system, how can the authorities be confident of meeting the needs of air pollution control? I feel it is imperative that the question of pollution control from unregistered premises should be considered, for failure to recognize the problems precludes the search for an answer. How can pollution from Britain's unregistered premises be controlled?

F. E. Ireland. We identify four questions from Mr Page, three of which, dealing with discovery of pollution incidents, monitoring and the control of air pollution, are interrelated.

The first question concerns levels of cadmium in the environment which constitute a health hazard. This should properly be addressed to the medical authorities. Nevertheless it can be said that no simple answer is possible since the particular sector of the environment, the route of exposure and the exposed population group will all influence the conclusion. However, in general, as far as human health is concerned, what matters is the *total* absorption of cadmium from all sources. For most people, food is the most important source and cigarette-smoking can be equally important. Respiration of airborne cadmium normally plays a very minor rôle in exposure. An expert working group convened by the E.E.C. has concluded that *continuous* exposure to a concentration of 2 $\mu g/m^3$ of cadmium in air might be expected to lead to kidney damage after 70 years exposure, 24 h a day. Long-term average values in the U.K. are normally less than one hundredth of this.

As far as cadmium in dust is concerned, no data are available on which to set hazardous levels. The group most likely to be at risk is very young children who might suck contaminated fingers or toys; clearly the degree of hazard will be a function of the individual child's behaviour. We know of no evidence of either acute or chronic health effects from this source of exposure.

To turn now to the other three questions, perhaps we can best illustrate how liaison between local authorities and the Alkali and Clean Air Inspectorate is helpful by referring to the 'incident' with which Mr Page was concerned. A programme of monitoring by the local authority indicated an undesirable degree of contamination by cadmium in a certain area, though its source was not unequivocally identified. The local authority consulted the Inspectorate who gave advice on the need for abatement equipment on the unregistered process operated at a nearby non-ferrous metal works and on the need for improvements in general housekeeping. Action was taken by the company to reduce emissions and to improve housekeeping.

This is clearly a perfectly proper way for matters to be handled when the local authority is responsible for control; the Alkali Inspectorate is always ready to help. In certain circumstances it has become apparent in the past that particular types of works caused an unacceptable degree of contamination and that local authorities have not had the necessary technical expertise to develop solutions. The Alkali Inspectorate has then assumed responsibility by scheduling the class of works under the Alkali, etc., Works Regulation Act 1906; an example here is mineral works where dust emissions were unacceptable.

The value of the liaison between local authorities and the Alkali Inspectorate was recognized in the Fifth Report of the Royal Commission on Environmental Pollution. Indeed, the Royal Commission considered that the present division of responsibilities should continue, with greater powers being afforded to the local authorities and more flexible arrangements for the transfer of control between central and local authorities. The Commission also proposed that local authorities should adopt a system of control by 'best practicable means', seeking guidance from the Alkali Inspectorate.

Finally, a word on monitoring. Monitoring surveys are carried out for a variety of reasons including, *inter alia,* prospective assessments of what is present and in what concentrations, specific checks on emissions from works and assessments of the efficiency of abatement equipment. Some major cities have carried out extensive programmes and a 'national network' of 20 sites is being operated by the Warren Spring Laboratory for a range of trace elements, in addition to the national survey of smoke and SO_2. Levels in country areas are also being assessed. But it remains a fact that most monitoring has been and will probably continue to be carried out specifically to look for certain substances which are expected, on the basis of local industry, to be present. This seems entirely sensible; it would be a waste of resources to monitor all parts of the country for a wide range of substances as a means of identifying sources of pollutants; it is far better in general to concentrate resources on dealing with areas expected or known to be contaminated. A comprehensive report on the present arrangements for monitoring our environment, *The monitoring of the environment in the United Kingdom*, Pollution Paper No. 1, has been prepared by the Department of the Environment and is available from Her Majesty's Stationery Office.

There are a few general points that I should like to make:

1. We can estimate that industry registered under the Alkali Act for England and Wales spends something of the order of £200M to £300M per year to meet the Inspectorate's requirements on air pollution control. We can guess that the remainder of industry probably spends a similar sum, giving a total for England and Wales of about £500M per year.

2. The Inspectorate has long appreciated the inadequacies and cost of spot-sampling of chimney emissions, especially for particulate matter. During the last decade it has sought to encourage the development of continuous monitoring instruments and has given research contracts to suitable organizations. Only by the use of continuous monitors can emissions be properly controlled by the owner.

3. The philosophy of the Alkali Inspectorate is to make industrial emissions to air safer than safe. Is 'best practicable means' good enough to achieve this end? I believe it is, and can exemplify this by our attitude towards airborne radioactive emissions, where we act as agents of the Secretary of State for the Environment under the Radioactive Substances Act 1960.

The International Commission on Radiological Protection has made recommendations on the doses of radioactivity to which the public can be exposed. From these can be calculated Derived Working Limits (d.w.l.) for individual factories. When 'best practicable means' are applied the actual working level of emission which is achieved is usually of the order of 1 % of the d.w.l. When incidents occur which raise the emission of radioactivity to 5–10 % of the d.w.l., the Inspectorate carries out an investigation to assess what caused the increase, with a view to preventing a recurrence. The same philosophy and policy is applied to works registered under the Alkali Act.